Ornamental Origami

Ornamental Origami

Exploring 3D Geomentric Designs

Meenakshi Mukerji

A K Peters, Ltd.
Wellesley, Massachusetts

Editorial, Sales, and Customer Service Office

A K Peters, Ltd.
888 Worcester Street, Suite 230
Wellesley, MA 02482
www.akpeters.com

Library of Congress Cataloging-in-Publication Data

Mukerji, Meenakshi, 1962–
        Ornamental origami : exploring 3D geometric designs / Meenakshi Mukerji.
            p. cm.
        Includes bibliographical references.
        ISBN 978-1-56881-445-2 (alk. paper)
        1. Origami. 2. Geometry in art. 3. Decoration and ornament. I. Title.
        TT870.M822 2008
        736'.982--dc22

                                                    2008060660

Cover Photographs

Front cover: QRSTUVWXYZ Stars

Back cover: Flowered Sonobe (top) and Icosahedron with Curves and Waves (bottom).

Printed in India

11 10 09 08   10 9 8 7 6 5 4 3 2 1

To my family, Pinaki, Ratul, and Rohit

Contents

Preface

I greatly enjoyed writing my first book *Marvelous Modular Origami* (2006, A K Peters, Ltd.). I was not intending to write another so soon; although I had a deluge of new origami designs flooding my mind by the time I had finished my first book. But so many readers inquired about a next book, that I was greatly motivated to author another. Here is the product of my imagination and your generous encouragement.

I learned a great deal from my first writing attempt and this book definitely reaps the benefits, now that I am not a novice author. This includes a better understanding of what makes an effective diagram, improved photography, greater attention to details, and an increased passion for doing a meticulous job overall. There are over 40 new models and lots of hints and clues for the folder, so you can make models that are wholly your own, or variations. For the mathematics lover, I have included the mathematical aspects of the design process, wherever possible.

The first chapter deals with origami basics and other material that has some overlap with *Marvelous Modular Origami*. Without this material, a modular origami book is perhaps not complete. In addition to origami symbols and bases, I have provided polyhedron charts and color distribution charts for reference during the assembly phase. Folding tips along with other matter important to folding origami, particularly modular origami, has been provided. The relationship between modular origami and mathematics has been discussed as well.

We start with simple models using *windmill bases* and *blintz bases* and then gradually progress into more intricate models such as the decorative icosahedra, the embellished Sonobe, and floral ball models, and conclude with a chapter on planar origami models. The planar model designs use a moderate amount of mathematics, some of which has been described in the chapter. A brief history and an exhaustive list of all planar models known to date, as compiled by renowned origami artist David Petty, has also been included. Some mathematical exercises have been provided at the conclusion of the book for those who would like to test their memory of any high school level mathematics.

Diagramming origami models is a bittersweet experience. Some units that are quite simple and may be taught easily in minutes in a face-to-face meeting can take hours to diagram. Easy moves may be difficult to illustrate using abstract diagrammatic methods. However, at the end of drawing a successful diagram there undoubtedly is great satisfaction and the "wrist-wrenching" experience of working the computer mouse extensively seems well worth the effort. A measure of any good origami diagram is that it should have practically no need for language. The universal international origami folding symbols that have evolved over time almost prove the language independence of this art. One can fold from diagrams explained in any language without having an iota of knowledge about the language itself. That is the beauty of the language of origami symbols.

This book presents a combination of simple to high intermediate models with an emphasis on the latter and should appeal to a wide range of audiences 12 years and older, with or without a mathematical background. I hope that you will enjoy this book as much as my previous one, if not more, and spread the joy of origami.

Cupertino, California

February 2008

Acknowledgments

This book would not have been possible without the encouragement that so many people have provided me and I would like to thank each and every one of them. Additionally, there are many people who have assisted me with the book by making direct or indirect contributions, and I will try my best to thank them one by one, hoping not to miss anybody. First of all, I would like to thank David Petty for contributing to the chapter on planar models a comprehensive history and a complete list of existing planar models including their finished model drawings. Next, I would like to thank Bennett Arnstein and Rona Gurkewitz for granting me permission to use their Triangle Edge Module [Gur95] in my Decorative Icosahedra chapter. Thanks to all those who have meticulously test folded, especially Rosalinda Sanchez, Halina Rosciszewska-Narloch, Rachel Katz, Aldo Marcell Velásquez, and Anne LaVin. No diagram is complete until test folded by somebody other than the author. I'd also like to thank Halina and Rosalinda for adding special effects to some of my photos. Thanks to Jean Jaiswal for proofreading all the text and for her enthusiasm. Thanks to Robert Lang for always being there to answer the various questions that I have had from time to time. Thanks to the Singhal family for their continued inspiration. Thanks to my publisher A K Peters, Ltd. for their tireless teamwork that refined my book. Thanks to all those who have folded models and contributed beautiful photographs. They have been acknowledged separately in the *Photo Credits* section. Thanks to David Lister, Thomas Hull, Paul Jackson, and Boaz Shuval for being readily available for consulting. Thanks to the numerous viewers of my website for providing constant encouragement. And finally, hearty thanks to my husband Pinaki, and sons Ratul and Rohit, for proofreading and providing all kinds of support and for simply being there.

Photo Credits

(In rough order of appearance)

- Bonefolder (page 3): Photo by Christine Cox of Volcano Arts http://www.volcanoarts.biz

- Chrysanthemum/Trillium (page 28) and Impatient/Gypsophila (page 45): Folding and photos by Aldo Marcell Velásquez (Nicaragua).

- Impatient (page 34) and 12-unit Chrysanthemum/Gypsophila (page 40): Folding by child artist Anjali Pemmaraju (California).

- Icosahedron with Curves 2 (pages 46 and 53): Folding by child artist Smriti Pramanick (California).

- Flowered Sonobe and Geranium (page 62), Dahlia Variation (page 73), Geranium Variation (page 78), Layered Petunia (page 82), and Daylily (page 86, top): Folding by Halina Rosciszewska-Narloch (Poland) and photos by Sebastian Janas (Poland).

- TUVWXYZ Stars (page 104): Folding and photo by Rosalinda Sanchez (Arizona).

- UVWXYZ Rectangles (page 113), STUVWXYZ Rectangles (page 124), and RTUVWXYZ Stars (page 127): Folding and photos by Matt Johnston (Washington).

- TUVWXYZ Rectangles (page 118) and STUVWXYZ Stars (page 122): Folding and photos by André Bracchi (France).

- All other folding and photos are by the author.

1 ◈ Modular Origami Basics

Perhaps most origami enthusiasts already know that the word *origami* is based on two Japanese words: *oru* (to fold) and *kami* (paper). Although this ancient art of paper folding started in Japan and China, origami is now a household word around the world. Most people have probably folded at least a paper boat or airplane in their childhood. Origami has now come a long way from traditional models and modular origami, origami sculptures, and tessellations are some of the newer forms.

The origin of modular origami is a little hazy due to the lack of proper documentation. It is generally believed to have taken off in the early 1970s with the Sonobe units made by Mitsunobu Sonobe, although it may have existed earlier. Six Sonobe units could be assembled into a cube. Three of those units could be assembled into a *Toshie Takahama Jewel* [Tak74] with one additional crease made to the units. With the additional crease direction reversed, Steve Krimball first formed the 30-unit ball [Gra76]. This dodecahedral-icosahedral formation, in my opinion, is the most valuable contribution to polyhedral modular origami. It gave birth to the idea that an unlimited number of origami models could be constructed based on various underlying polyhedra. Later on, Kunihiko Kasahara, Tomoko Fuse, Miyuki Kawamura, Lewis Simon, Bennet Arnstein, Rona Gurkewitz, David Mitchell, Francis Ow, and many others made significant contributions to modular origami. Thomas Hull and Robert Lang added the notable family of the highly mathematical polypolyhedra (a term coined by Lang) that are essentially interwoven polygonal or polyhedral frames [Hul02]. As Lang notes, these polypolyhedra have an "uncanny" beauty [Lan] and personally, I find them extremely challenging and enjoyable to make.

Modular origami, as the name suggests, involves assembling several usually identical modules or units to form one finished model. Modular origami almost always implies polyhedral or geometric modular origami, although there are a number of other modular models that have nothing to do with polyhedra. Dinosaur skeletons made of several square pieces of paper and a traditional Chinese modular unit with which one can virtually construct any form are but a few examples. Generally speaking, glue is not required, but for some models it is recommended for increased longevity, and for some others glue might be essential simply to hold the units together. The models presented in this book do not require any glue except for when they are intended to be rough handled.

The symmetry of modular origami models is appealing to almost everyone, especially to those who have an appreciation for polyhedra. While an understanding of mathematics is useful for designing these models, it is not crucial for merely following polyhedra charts and instructions to construct the models. The beauty lies in the fact that even though mathematics is not one's forte, one can still construct these models and perhaps even have a different appreciation of the mathematical principles involved. Like any multi-stepped task that requires patience and diligence, the end result of one's hard work is a reward well earned and enjoyed. Aesthetics and mathematics brilliantly come together in these wonderful origami structures.

Modular origami can be fit relatively easily into one's busy schedule. Unlike other art forms, one does not need a long uninterrupted stretch of time all at once. Upon mastering a unit that takes very little time, batches of it can be folded anywhere anytime, including the very short free periods that one might have in between other work. When the units are all folded, the assembly can also be done slowly over time. I have been a working mother with two young boys and I am quite aware of how limited free time can be. But as I have found out, modular origami can easily trickle into the nooks and crannies of one's packed day without jeopardizing much else. Those long waits at the doctor's office or anywhere else and those long rides or flights do not have to be boring and unproductive any longer. Just remember to carry some paper and diagrams along and you are ready with practically no extra baggage.

Folding Tips and Tools

◈ Use paper of the same thickness and texture for all units that make up a model. This ensures that the finished model will hold evenly and look symmetric. Virtually any paper from color photocopy paper to gift wrap works. The origami paper commonly available on the market that is colored on one side and white on the other, usually referred to as *kami*, works for most models. There is a host of other fancy papers available: foil-backed paper, duo, *washi*, *chiyogami*, elephant hyde paper, etc.

◈ Pay attention to the grain of the paper. Make sure that when starting to fold, the grain of the paper is oriented the same way for all units. This is important so as to ensure uniformity and homogeneity of the model. To determine the grain of the paper, gently bend paper both horizontally and vertically. The grain of the paper is said to lie along the direction that offers less resistance during bending.

◈ Accuracy is particularly crucial to modular origami, so your folds need to be as accurate as possible. Only then will the finished models look symmetric and neat.

◈ It is advisable to fold a trial unit before folding the real units. This gives you an idea of the finished unit size. In some models the finished unit is much smaller than the starting paper size, while in others this is not so. Making a trial unit will give you an idea of what the size of the finished units— and hence a finished model—might be, starting with a certain paper size. It will also give you an idea about the paper properties and whether the paper type selected is suitable for the model.

◈ After you have determined your paper size and type, procure ALL the paper you need for the model before starting. If you do not have all the paper at the beginning, you may find, as has been my experience, that you are not able to find more paper of the same kind to finish your model.

◈ If a step looks difficult, looking ahead to the next step often helps immensely. This is because the execution of a current step results in what is diagrammed in the next step.

◈ Assembly aids such as miniature clothespins or paper clips are often advisable, especially for beginners. Some assemblies simply need them whether you are a beginner or not. These pins or clips may be removed as the assembly progresses or upon completion of the model.

◈ During assembly, putting together the last few units, especially the very last one, can be challenging. During those times, remember that it is paper you are working with and not metal! Paper is flexible and can be bent or flexed for ease of assembly.

◈ After completion, hold the model in both hands and compress gently to make sure that all the tabs are securely and completely into their corresponding pockets. Finish by working around the ball.

◈ Many units involve folding into thirds. The best way to do this is to make a template using the same size paper as the units. Fold the template into thirds using the method explained in the *Origami Symbols and Bases* section of this chapter. Then use the template to crease your units. This saves time and reduces unwanted creases.

◈ Procure the minimal basic handy tools listed on the next page. These tools assist in sizing paper, making neat crisp creases, curling paper (used extensively in this book), and assembling models.

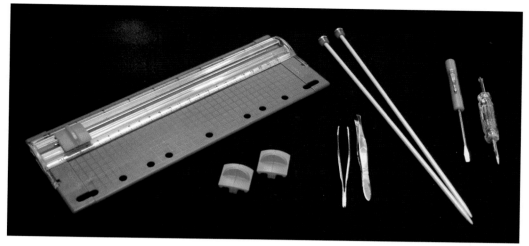

Left to right: Portable photo trimmer with replaceable blades are great for trimming origami paper to size. Tweezers are useful for accessing hard-to-reach places. Knitting needles, screwdrivers, or similar objects, such as narrow pencils work well for curling paper (used extensively in this book).

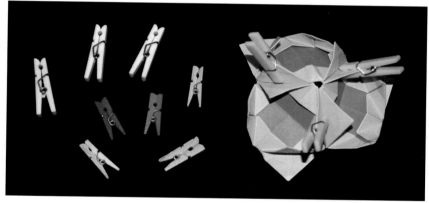

Miniature clothespins may be used during model assembly as temporary aids to hold two adjacent units together. The clothespins may be removed as the assembly progresses or after completion.

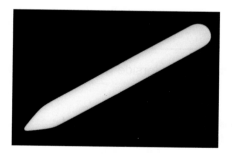 

A traditional bone folder (left) is used for making neat and crisp folds. Plastic milk jug handle cutouts, often found at grocers' refrigerator (with some luck) or objects such as discarded credit cards work almost as well.

Origami Symbols and Bases

This is a list of commonly used origami symbols and bases. While it covers all symbols and bases referenced in this book, it is by no means a complete list.

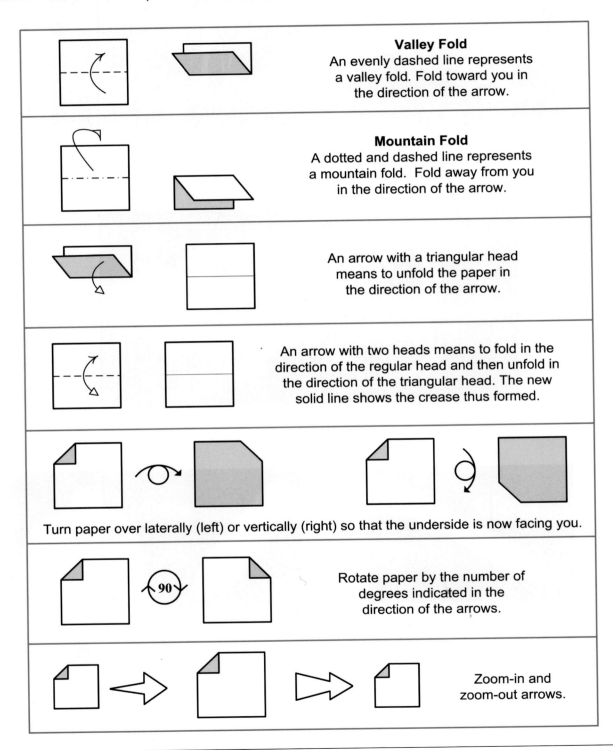

Valley Fold
An evenly dashed line represents a valley fold. Fold toward you in the direction of the arrow.

Mountain Fold
A dotted and dashed line represents a mountain fold. Fold away from you in the direction of the arrow.

An arrow with a triangular head means to unfold the paper in the direction of the arrow.

An arrow with two heads means to fold in the direction of the regular head and then unfold in the direction of the triangular head. The new solid line shows the crease thus formed.

Turn paper over laterally (left) or vertically (right) so that the underside is now facing you.

Rotate paper by the number of degrees indicated in the direction of the arrows.

Zoom-in and zoom-out arrows.

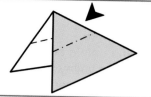 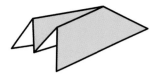

**Reverse Fold or
Inside Reverse Fold**
Push in the direction of the arrow to arrive at the result.

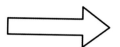 

Pull out paper.　　　　Equal lengths.　　　　Equal angles.

Figure is truncated for diagramming convenience.

Repeat once, twice, or as many times as indicated by the tail of the arrow.

Fold from dot to dot with the circled point as pivot.

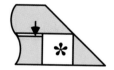 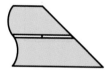

✱ : Tuck in opening underneath.

Fold repeatedly to arrive at the result.

  Pleat Fold

An alternate mountain and valley fold to form a pleat. Two examples are shown.

 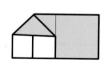

Squash Fold
Turn paper to the right along the valley fold while making the mountain crease such that *A* finally lies on *B*.

Cupboard Fold
First fold and unfold the centerfold, also called the *book-fold*, then valley fold the left and right edges to the center like cupboard doors.

Blintz Base

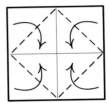 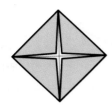

Valley fold and unfold both book-folds. Then valley fold all four corners to the center.

Windmill Base

Please see the beginning of the next chapter.

Waterbomb Base

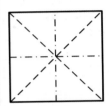 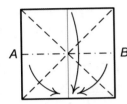 

Valley fold and unfold diagonals, then mountain fold and unfold book-folds. 'Break' line *AB* at the center and collapse such that *A* meets *B*.

Preliminary Base

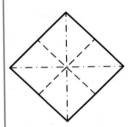 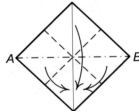

This is similar to the waterbomb base above, but the mountain and valley folds are reversed, i.e., the diagonals are mountain folded and the book-folds are valley folded at start.

Petal Fold

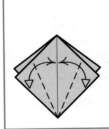 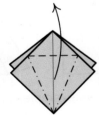  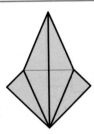

The figure on the left illustrates petal folding on a flap of a preliminary base.

Folding a Square into Thirds

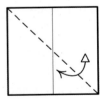 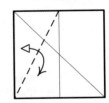 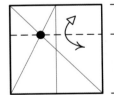 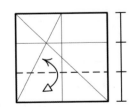

1/3

2/3

Fold and unfold book-fold and one diagonal. Then fold and unfold diagonal of one rectangle to find 1/3 point. Fold and unfold the bottom rectangle into half.

Modular Origami and Mathematics

"Mathematicians often rhapsodize about the austere elegance of a well-wrought proof. But math also has a simpler sort of beauty that is perhaps easier to appreciate: It can be used to create objects that are just plain pretty—and fascinating to boot." This statement by Julie J. Rehmeyer [Reh08] echoes my own sentiments. In my opinion, polyhedral modular origami is an inspiring example of how mathematics, art, and beauty all come together in a wonderful cross-pollination of disciplines that all can understand and appreciate without the rigors of a proof. It perfectly demonstrates the balance and symmetry often found in mathematics in a hands-on intuitively gripping way and can be compelling to mathematicians and artists alike. A mathematician's delight in pondering the symmetry of the polyhedra underlying the models is just a different spin on the same delight an artist experiences when contemplating the beauty of the exact same object. Human beings instinctively find symmetry pleasurable and beautiful—studies have even linked symmetry to our perception of beauty in facial features, though it is not quite relevant here, it is worth mentioning.

Although most people view origami as child's play and libraries and bookstores tend to shelve origami books in the juvenile section, there is much more to origami than generally perceived. There is such a primal connection between mathematics and origami that a separate international origami conference, *Origami Science, Math, and Education* (OSME) was spun off in 1989. Professor Kazuo Haga of the University of Tsukuba, Japan, appropriately proposed the term *Origamics* at 3OSME in 1994 [Hul02] to refer to this genre of origami that is heavily related to science and mathematics. The mathematics of origami has been extensively studied not only by origami enthusiasts but also by many mathematicians, scientists, and engineers as well as artists. These interests date back to at least the late nineteenth century when Tandalam Sundara Row of India wrote the book *Geometric*

Exercises in Paper Folding in 1893 [Row66]. He used novel methods to teach concepts in Euclidean geometry by simply using scraps of paper and a penknife. The geometric results were so easily attainable that it inspired quite a few other mathematicians to investigate the geometry of paper folding for the first time.

As mentioned in the introduction, modular origami almost always implies polyhedral or geometric modular origami, although there are some exceptions. A polyhedron is a three-dimensional solid that is bound by polygonal faces. A polygon, in turn, is a two-dimensional figure bound by straight lines. Aside from the polyhedra themselves, modular origami involves construction of a host of other objects that are based on the principles of polyhedra. Looking at the external artistic, often floral or ornamental appearance of modular origami models, it is hard to believe that there is any mathematics at all involved. But in fact, for every model there is a hidden underlying polyhedron. It could be based on the Platonic or Archimedean solids listed in the next section, or it could be based on prisms, antiprisms, Kepler-Poinsot solids, Johnson's Solids [Joh66], rhombohedra, symmetrohedra [Kap01], or even irregular polyhedra. The most referenced polyhedra for origami constructions are undoubtedly the five Platonic solids, followed by some of the Archimedean solids.

Assembly of the units that comprise a model may at first seem very puzzling to the novice, or even downright impossible. But understanding certain mathematical aspects can considerably simplify the process. First, one must determine whether a unit is a face unit, an edge unit, or a vertex unit, i.e., whether a unit identifies with a face, an edge or a vertex respectively, of the underlying polyhedron. Face units are the easiest to identify. The windmill base models presented in the next chapter are face units. There are only a few known vertex units, an example being David Mitchell's Electra [Mit00]. Most modular model units and all the rest of the models presented in this book are edge units. For edge units there is a second step involved—one must identify which part of the unit, which is far from looking like an edge, actually translates to the edge of a polyhedron. Although it may appear perplexing at first, on closer look one may find that it is not an impossible task. Once the identifications are made and the folder can see through the maze of superficial designs and perceive the unit as a face, an edge, or a vertex, assembly becomes simple. It is then just a matter of following the structure of the underlying polyhedron to assemble the units. With enough practice, even the polyhedron chart need not be consulted anymore.

When designing a model, one might start with a mathematical approach. Alternatively, one might create a model intuitively and later open it back out into the flat sheet of paper it originated from and then ponder the mathematics of the creases. Further refinement of the design, if required or desired, can be achieved by studying the creases. Either method of design is perfectly valid. Some mathematical approach to designing has been loosely illustrated in the last chapter, *Planar Mod-*

els. Various desired angles arrived at using origami methods have been explained and proofs for the same have been provided.

Modular origami is also a great tool for studying and teaching concepts in three-dimensional geometry. It is a wonderful engaging hands-on companion to teaching theoretical concepts. While it is one thing to look at pictures of polyhedra, it is quite another thing to construct one. That origami simply requires easy methods and inexpensive materials accessible to all, lends itself perfectly to the classroom. Nothing can compare to the learning achieved when actually holding one's created polyhedron and studying it both during and after construction. It can make the dull or confusing mathematical principles leap to life from the page. The spatial relationship among edges and vertices and the duality among polyhedra stand out clearly. Origami is also a wonderful tool for constructing and studying crystal structures in physical chemistry classes. Presented next are polyhedra and color distribution charts that usually turn out to be handy for polyhedral modular origami constructions.

Platonic and Archimedean Solids

Below is a list of Platonic and Archimedean solids commonly referenced for origami constructions.

Platonic Solids

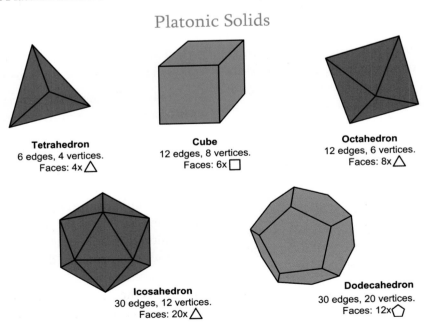

Tetrahedron
6 edges, 4 vertices.
Faces: 4x △

Cube
12 edges, 8 vertices.
Faces: 6x □

Octahedron
12 edges, 6 vertices.
Faces: 8x △

Icosahedron
30 edges, 12 vertices.
Faces: 20x △

Dodecahedron
30 edges, 20 vertices.
Faces: 12x ⬠

Modular Origami Basics

Cuboctahedron
24 edges, 12 vertices.

Faces:

8x △

6x ☐

Truncated Octahedron
36 edges, 24 vertices.

Faces:

6x ☐

8x ⬡

Rhombicuboctahedron
48 edges, 24 vertices.

Faces:

8x △

18x ☐

Truncated Cuboctahedron
72 edges, 48 vertices.

Faces:

12x ☐

8x ⬡

6x ⬡

Icosidodecahedron
60 edges, 30 vertices.

Faces:

20x △

12x ⬠

Truncated Icosahedron
90 edges, 60 vertices.

Faces:

12x ⬠

20x ⬡

Rhombicosidodecahedron
120 edges, 60 vertices.

Faces:

20x △

30x ☐

12x ⬠

Snub Cube
60 edges, 24 vertices.

Faces:

32x △

6x ☐

Note that five of the thirteen Archimedean Solids have not been shown: truncated tetrahedron, truncated cube, truncated cuboctahedron, truncated icosidodecahedron, and snub dodecahedron.

Even Color Distribution

This section illustrates a few polyhedra with even color distribution for their edges. For most modular origami constructions, each edge of a polyhedron maps to a module or unit.

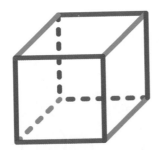

Three-color tiling of a cube
(every vertex has three distinct colors)

Four-color tiling of a cube
(every face has four distinct colors)

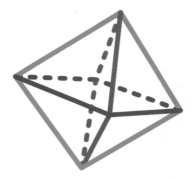

Three-color tiling of an octahedron
(every face has three distinct colors)

Four-color tiling of an octahedron
(every vertex has four distinct colors)

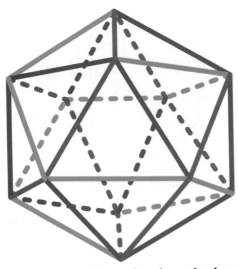

Three-color tiling of an icosahedron
(every face has three distinct colors)

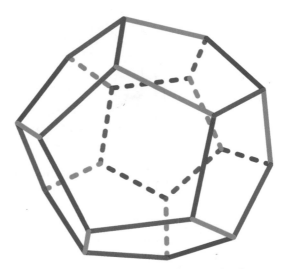

Three-color tiling of a dodecahedron
(every vertex has three distinct colors)

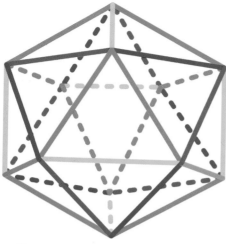

Five-color tiling of an icosahedron
(every vertex has five distinct colors and
every face has three distinct colors)

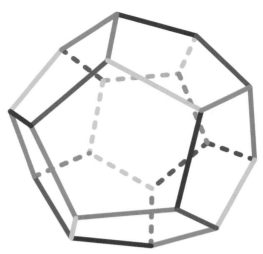

Five-color tiling of a dodecahedron
(every face has five distinct colors and
every vertex has three distinct colors)

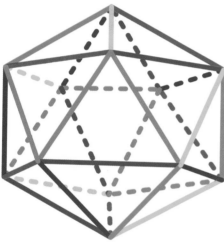

Six-color tiling of an icosahedron
(every vertex has five distinct colors and
every face has three distinct colors)

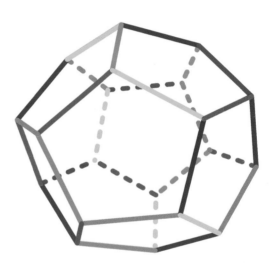

Six-color tiling of a dodecahedron
(every face has five distinct colors and
every vertex has three distinct colors)

It is implicit that for an even color distribution, the number of colors one chooses must be a factor of the number of edges or units in a model. For example, for a 30-unit model one can use three, five, six or ten colors. In all of these figures each edge represents one origami unit. The dotted lines are invisible from the point of view. The number of units you need to fold per color is equal to the total number of edges divided by the total number of colors. Even color distribution makes a model more appealing but for some models the use of a single color is more effective while, for some others random coloring works equally well.

Tuberoses and Oleanders (top) and Tuberose, standalone and in a ball (bottom).

Windmill Base Models

2 ◆ Windmill Base Models

The models in this chapter are based on a traditional base called the *windmill base*. The windmill base is very versatile and many different models can be folded from it. The traditional Chrysanthemum Kusudama and a host of other kusudamas, including those by Friedrich Froebel, Tomoko Fuse [Fus02] and Kunihiko Kasahara [Kas03] begin with

windmill bases. Shown below are instructions on how to fold a windmill base. They will be used throughout the chapter. At the end of the chapter, we will make a model with a pentagonal version of the windmill base. All paper types including photocopy paper are suitable for these models and squares of sides 3″–5″ are recommended.

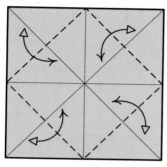

1. Valley fold both diagonals and book-folds and unfold, then fold corners to center and unfold.

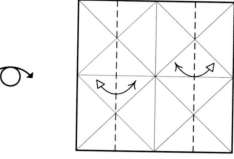

2. Cupboard fold and unfold.

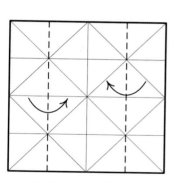

3. Cupboard fold.

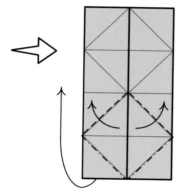

4. Make the mountain and valley folds as shown while bringing bottom edge to center.

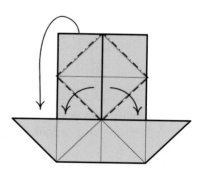

5. Repeat step 4 on the top, bringing top edge to center.

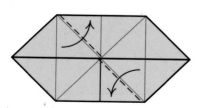

6. Valley fold the two flaps in the directions indicated.

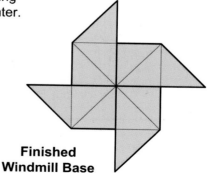

Finished Windmill Base

Sunburst

Start with a windmill base as diagrammed on page 13 and then continue with the following steps.

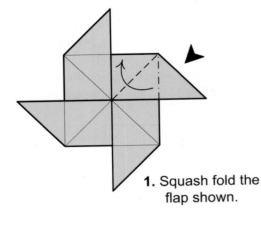

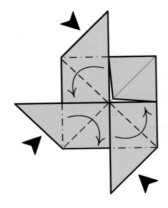

1. Squash fold the flap shown.

2. Squash fold the other three flaps.

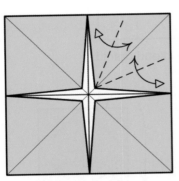

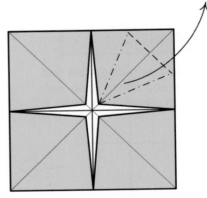

3. Valley fold edges to diagonal and unfold.

4. Valley and mountain fold taking the corner outwards. This is also called *petal fold*.

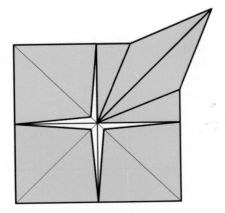

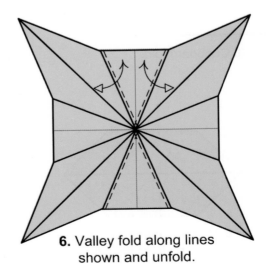

5. Repeat Steps 3 and 4 on the other three corners.

6. Valley fold along lines shown and unfold.

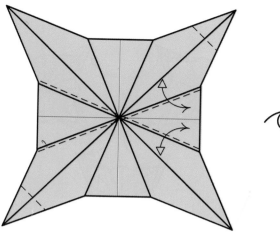

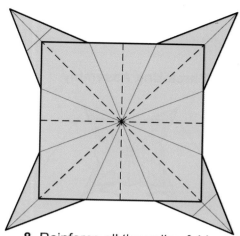

7. Valley fold along lines shown and unfold. Also, crease two diagonally opposite corners to mark tabs.

8. Reinforce all the valley folds along the diagonals and book-folds within the square through all layers. Push center down to finish unit.

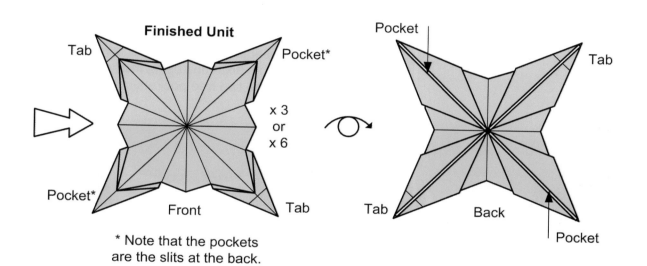

Finished Unit

Tab

Pocket*

x 3
or
x 6

Pocket*

Front

Tab

* Note that the pockets are the slits at the back.

Pocket

Tab

Tab

Back

Pocket

Assembly

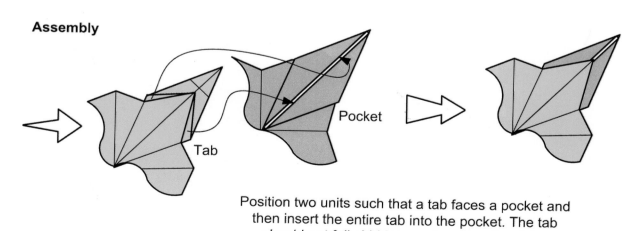

Tab

Pocket

Position two units such that a tab faces a pocket and then insert the entire tab into the pocket. The tab should get fully hidden as shown on the right.

Three-Unit Assembly

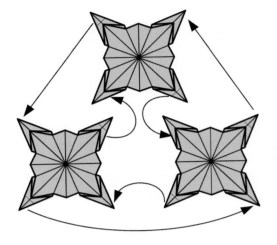

Six-Unit Assembly

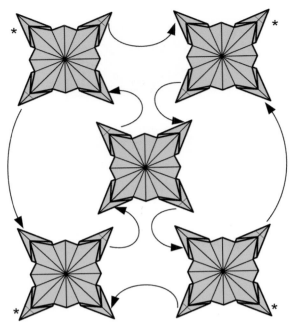

Assemble five units as shown above.
Attach a sixth unit using the two free tabs
and the two free pockets marked with *.

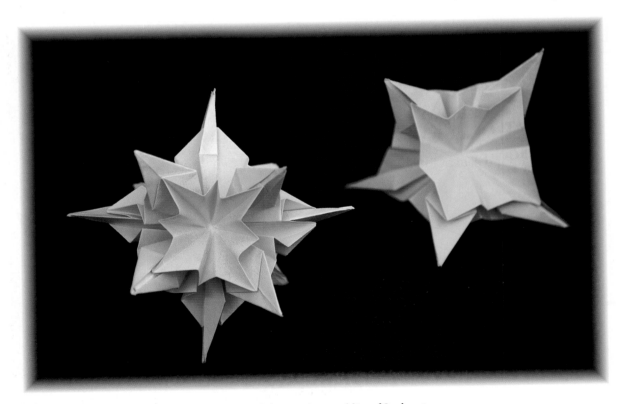

Six-unit and three-unit assemblies of Sunburst.

Windmill Base Models

Starburst

Start with a windmill base as diagrammed on page 13 and then continue as follows:

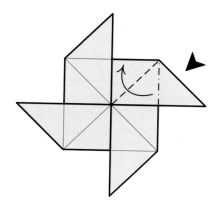 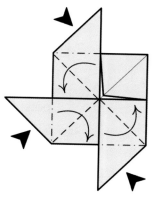

1. Squash fold the flap shown.

2. Squash fold the other three flaps.

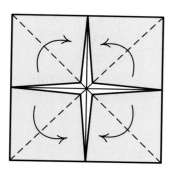

3. Valley fold the four flaps as shown.

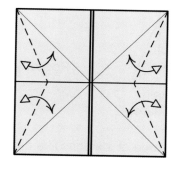

4. Valley fold as shown and unfold.

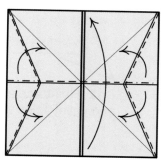

5. Collapse as shown to bring the bottom edge to the top.

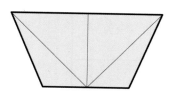

6. Rotate 90°.

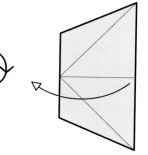

7. Unfold last step.

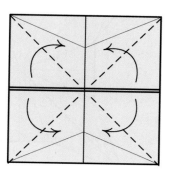

8. Valley fold flaps as shown.

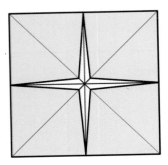

9. Repeat Steps 3–8.

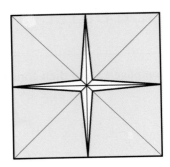

10. Repeat steps 3–5 of Sunburst model to petal fold corners.

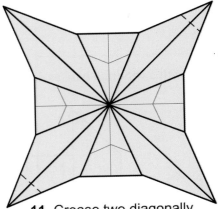

11. Crease two diagonally opposite corners to mark tabs.

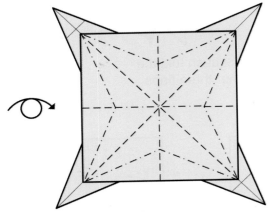

12. Reinforce all the mountain and valley folds within the square.

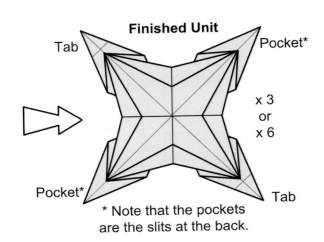

Finished Unit

Tab

Pocket*

x 3 or x 6

Pocket*

Tab

* Note that the pockets are the slits at the back.

Assemble three or six units like the Sunburst model.

Starburst Unit Variation
Do Steps 1–5 of the Sunburst model.

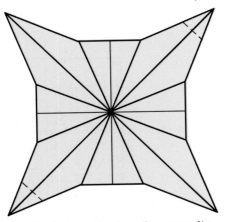

1. Mark two diagonally opposite corners for tabs and turn over.

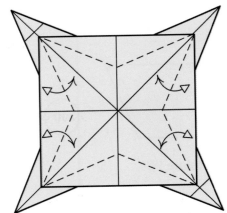

2. Valley fold and open to bisect the 45° angles.

Windmill Base Models

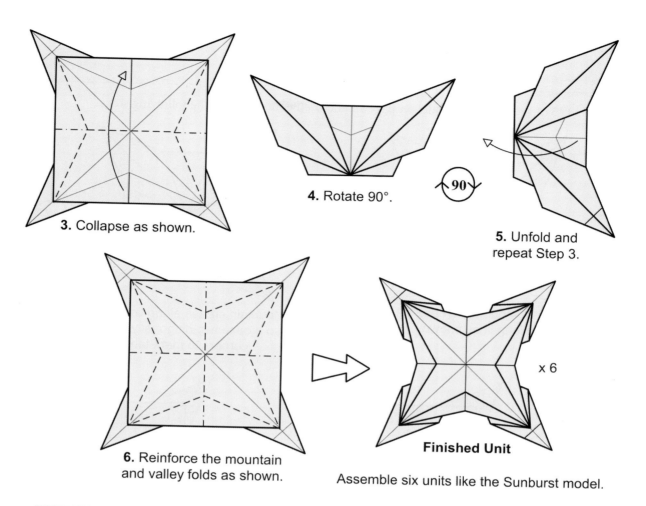

3. Collapse as shown.

4. Rotate 90°.

⟨**90**⟩

5. Unfold and repeat Step 3.

6. Reinforce the mountain and valley folds as shown.

Finished Unit

Assemble six units like the Sunburst model.

x 6

Three-unit assembly of Starburst and six-unit assembly of Starburst variation.

Ixora

Do Steps 1–5 of the Sunburst model on page 14 and then continue with the following steps.

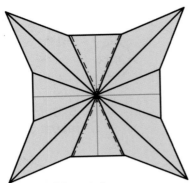

1. Mountain crease along lines shown.

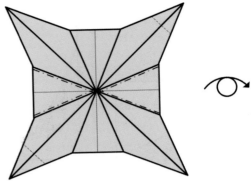

2. Mountain crease along lines shown. Also crease two diagonally opposite corners to mark tabs.

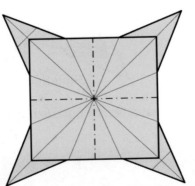

3. Reinforce the book-folds as mountain creases through all layers.

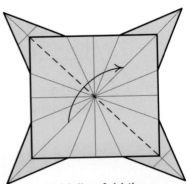

4. Valley fold the diagonal without folding the two tabs behind.

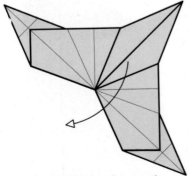

5. Unfold and repeat the other diagonal.

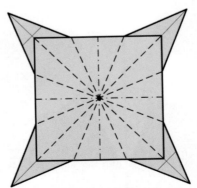

6. Dip down the center of the square and reinforce mountain and valley folds to arrive at the finished unit.

Finished Unit

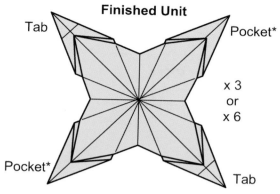

Tab   Pocket*

x 3
or
x 6

Pocket*   Tab

* Note that the pockets
are the slits at the back.

Assemble three or six units
like the Sunburst model
presented in the beginning
of this chapter.

Six-unit and three-unit assemblies of Ixora.

Tuberose

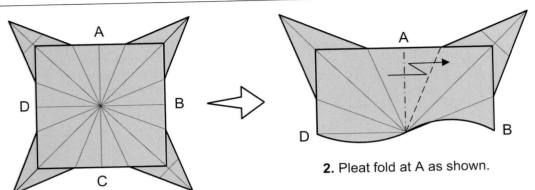

1. Start with a finished Ixora unit, flattened.

2. Pleat fold at A as shown.

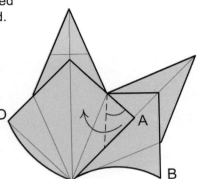

3. Fold tip A as shown such that the marked angle is approximately 45° or a tad bit larger. Crease firmly.

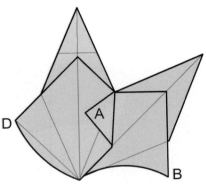

4. Repeat Steps 2 and 3 at B, C, and D. Earlier folds tend to unfold; repeat folding if necessary until all four corners stay folded.

Finished Unit

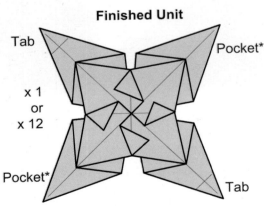

Tab

Pocket*

x 1
or
x 12

Pocket*

Tab

* Note that the pockets are the slits at the back.

Assembly

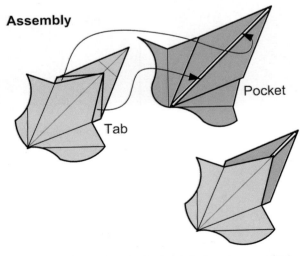

Tab

Pocket

Position two units such that a tab faces a pocket and then insert the entire tab into the pocket. The tab should be fully hidden as shown on the right.

To make a standalone Tuberose, make a unit with tabs unmarked and curl the four petals with a narrow pencil-like object.

Twelve-Unit Assembly

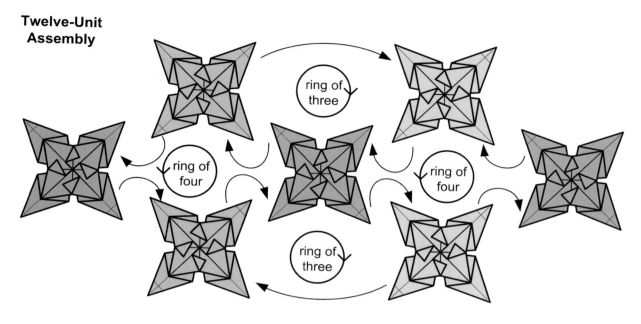

To make a 12-unit ball, assemble as above. Assembly is shown here in a three-color scheme. Assemble seven units as shown starting with the center unit.

Assemble the rest of the units such that for each and every unit, two three-unit rings and two four-unit rings lie diametrically opposite to each other.

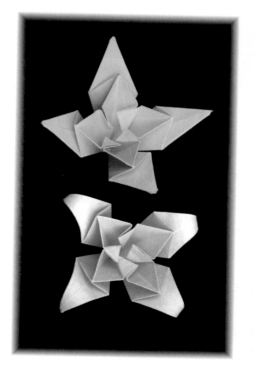

Tuberoses, standalone (left), and as a 12-unit ball (right).

Oleander

The Oleander model is an extrapolation of the Tuberose model on to pentagonal paper. Folding with pentagonal paper is not uncommon in origami. Philip Shen, Rona Gurkewitz, Bennett Arnstein, and several other origami artists have models folded from pentagons. Once, while playing with a pentagonal piece of paper, I mapped the folds of a traditional Lily/Iris from a square on to a pentagon and the resulting flower was very beautiful and more life like. There are quite a few origami models that can be translated from a square to a pentagon, leading to pleasing end results.

There exist several techniques for obtaining a pentagon starting with a square piece of paper. Below is a widely used method [Kawai70] that I find easy and elegant with minimal wastage of paper. The resulting pentagon is not perfect but the error is minimal and it works perfectly well for regular scale origami purposes.

A Traditional Method for Obtaining a Pentagon from a Square

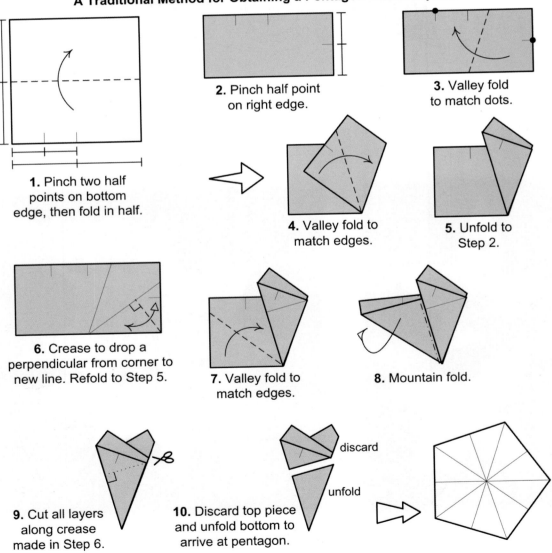

1. Pinch two half points on bottom edge, then fold in half.

2. Pinch half point on right edge.

3. Valley fold to match dots.

4. Valley fold to match edges.

5. Unfold to Step 2.

6. Crease to drop a perpendicular from corner to new line. Refold to Step 5.

7. Valley fold to match edges.

8. Mountain fold.

9. Cut all layers along crease made in Step 6.

10. Discard top piece and unfold bottom to arrive at pentagon.

discard

unfold

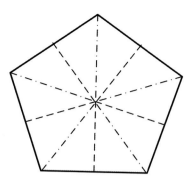

1. Mountain and valley fold along pre-existing creases as shown.

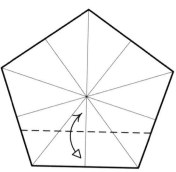

2. Bring bottom edge to center and crease and unfold.

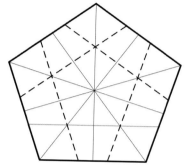

3. Repeat Step 2 on all other edges.

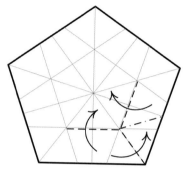

4. Valley and mountain fold as shown.

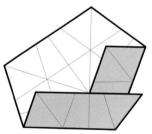

5. Repeat Step 4 on all other corners.

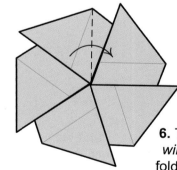

6. This is a *pentagonal windmill base*. Valley fold top flap to the right.

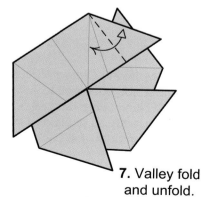

7. Valley fold and unfold.

8. Squash.

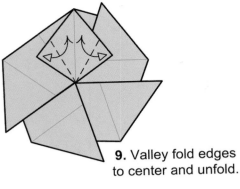

9. Valley fold edges to center and unfold.

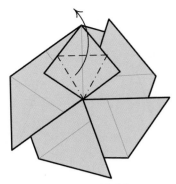

10. Petal fold.

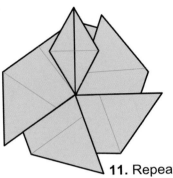

11. Repeat Steps 6-10 on all corners.

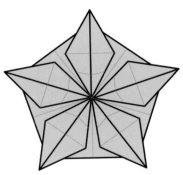

12. Turn over.

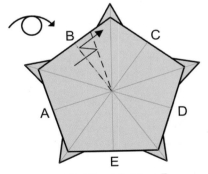

13. Pleat fold at B as shown.

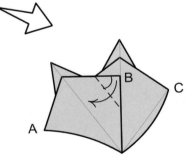

14. Fold tip B as shown such that the marked angle is approximately 60°. Crease firmly.

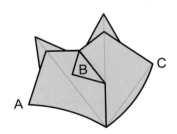

15. Repeat Steps 13 and 14 at C, D, E, and A. Earlier folds tend to unfold; repeat folding if necessary until all five corners stay folded.

Finished Unit   x 1 or x 12

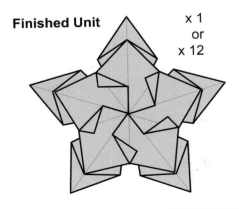

For a standalone Oleander, curl all ten outer and inner tips outwards to complete flower. For a 12-unit Oleander ball, lock tabs and pockets exactly as explained in the previous model, the Tuberose.

Assembly

Assemble with each unit lying along the face of a dodecahedron and at each vertex three units locking in a ring as shown. Since the total number of tabs and pockets per unit is odd (five), the tab-pocket distribution per unit will not be even across the model.

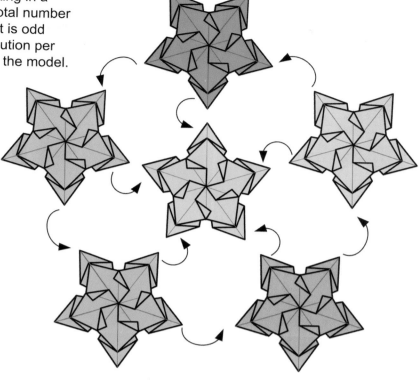

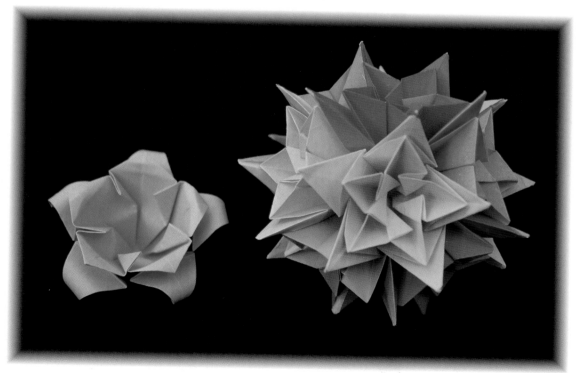

Oleander, standalone (left) and in a 12-unit ball.

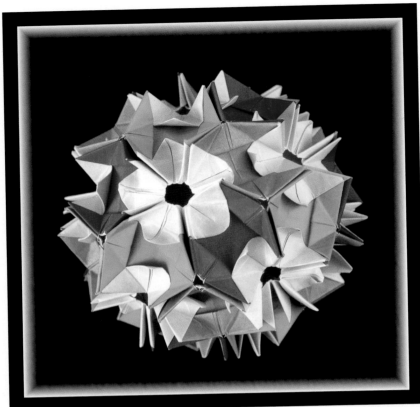

Periwinkle (top) and Chrysanthemum/Trillium in duo paper (bottom), both without inserts.

Blintz Base Bouquets

3 ◆ Blintz Base Bouquets

While playing with *blintz bases* I came up with a bunch of flower bouquets that are diagrammed in this chapter. Blintz base modular origami models are not new and some other examples that come to mind are Donatella Cecconi's Petali [Cecc89] and Mio Tsugawa's Mosaic Box [Tsu] of 2003. These models all utilize repeated blintz folding.

As with most other modular models, the units can be assembled into bouquets of 6, 12, or 30 units; the 30-unit ones being the most attractive because their five petal flowers seem more natural. We begin with making the Basic Blintz Unit models. Then we will look into the possible variations achieved by forming different flowers at the four or five-point vertices and the three-point vertices, which are actually the faces of the octahedral or icosahedral assemblies. We can form the following flowers at the four or five-point vertices: Impatients, Periwinkles, and Chrysanthemums. At the three-point vertices or faces, we can form Trilliums or Gypsophilas or simply leave them plain, thus accentuating the flowers at the five-point vertices. Different combinations of flowers lead to many different finished models or bouquets, e.g., Chrysanthemum, Chrysanthemum Trillium, Chrysanthemum Gypsophila, Impatient, Impa-tient Trillium, Impatient Gypsophila, Periwinkle, Periwinkle Trillium, and Periwinkle Gypsophila.

To further enhance the beauty of the models, use contrasting colors between the flowers and the background of the models. This can be achieved by using a simple insert in every unit. Another way to obtain contrast is to use harmony paper. There is a plethora of harmony papers available these days, the use of which can yield dramatic end results. Unlike regular *kami*, which is one solid color, harmony paper may have various colors blending into one another on one sheet or various shades of the same color on one sheet. A third way to achieve contrast would be to use regular *kami* and perform some additional steps to get the desired color contrast. This method, discussed at the end of the chapter, would be a purist's choice, but the finished units get one layer thicker due to the additional folding. The thickness is manageable and not a problem. For this no-insert method, duo paper can be very effective.

A note on assembly—the models are sturdy without glue but during assembly, aids such as miniature clothespins are recommended, especially for beginners. The clothespins can be removed as the assembly progresses or at the end.

Recommendations

Paper Size: 3″–5″ squares

Paper Type: *Kami* or harmony paper

Finished Model Size

3″ squares yield 30-unit models that are approximately 3.75″ in height

Basic Blintz Unit

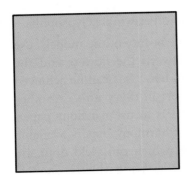

Paper for main unit.

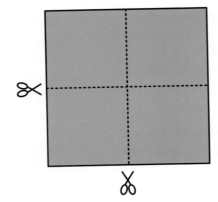

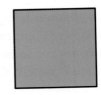

Use contrasting color for inserts and cut into quarters.

Paper for inserts (optional).

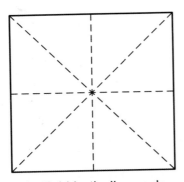

1. Fold both diagonals and book-folds and unfold.

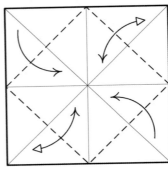

2. Blintz fold the two corners shown and unfold.

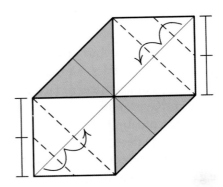

3. Valley fold each corner twice.

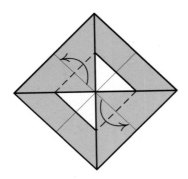

4. Valley fold corners.

5. Turn over.

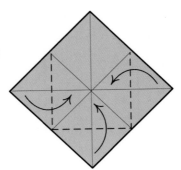

6. Blintz fold three corners shown.

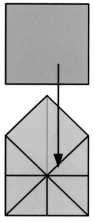

7. Slide optional insert all the way into the "envelope."

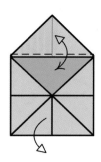

8. Valley fold top corner and unfold. Unfold bottom corner.

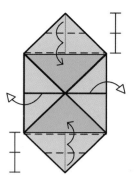

9. Valley fold two corners twice as shown, then unfold the other two blintz folds.

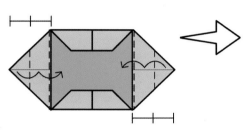

10. Valley fold the other two corners twice.

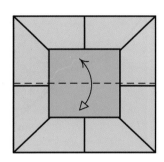

11. Valley fold in half and unfold. (Skip this step for a six-unit assembly.)

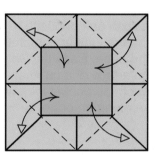

12. Blintz fold firmly through all layers and unfold.

Finished Unit

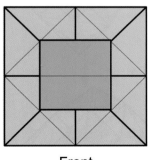

Front

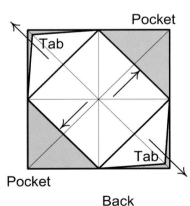

Back

x 6 or x 12 or x 30

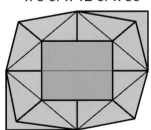

Reinforce the blintz folds from Step 12. The finished unit now becomes three-dimensional.

Six-Unit Assembly

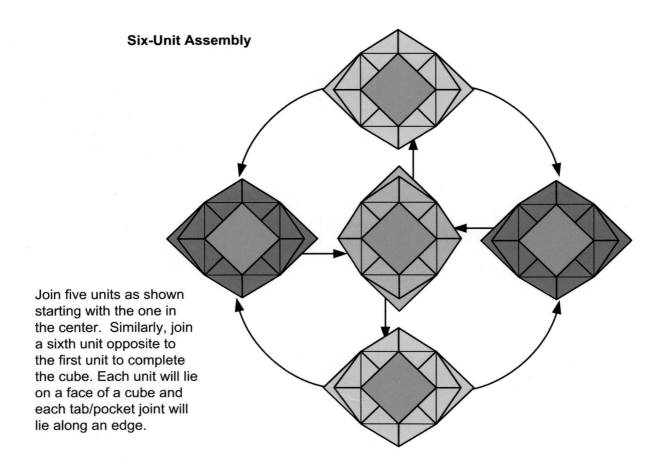

Join five units as shown starting with the one in the center. Similarly, join a sixth unit opposite to the first unit to complete the cube. Each unit will lie on a face of a cube and each tab/pocket joint will lie along an edge.

Twelve-Unit Assembly

Assemble four units in a ring to form a vertex of an octahedron. Add a fifth unit (shown in purple) to form a face. Continue assembling in octahedral symmetry to complete the model.

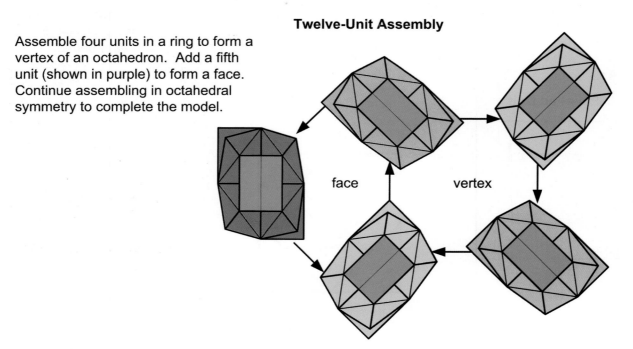

face   vertex

Note that for 12-unit and 30-unit assemblies, Step 9 is mandatory. The crease made in step 9 will lie along the edges of the assembled polyhedron.

Thirty-Unit Assembly

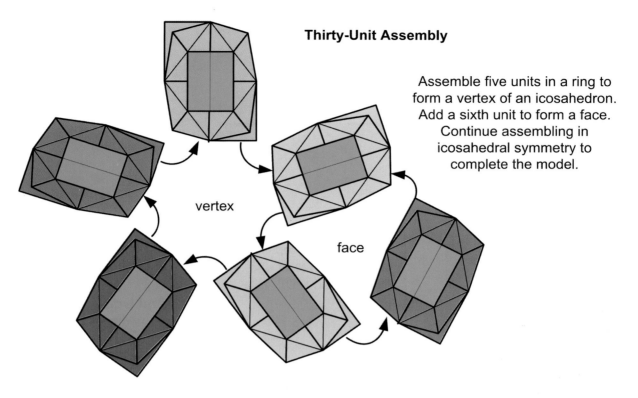

Assemble five units in a ring to form a vertex of an icosahedron. Add a sixth unit to form a face. Continue assembling in icosahedral symmetry to complete the model.

vertex

face

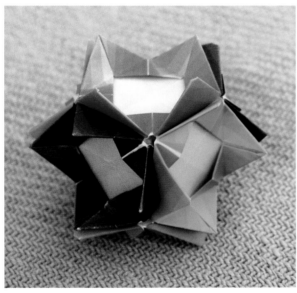

Assemblies of the Basic Blintz Unit: six-unit assembly of harmony paper (left) and 12 unit assembly of *kami* with gold inserts.

Impatient

Start with a completed and flattened Basic Blintz Unit.

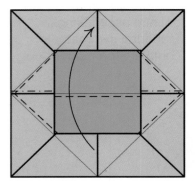

1. Bring bottom edge to top while making the mountain and valley creases as shown.

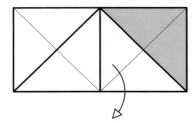

2. Unfold last step gently to arrive at the finished unit.

Gently reinforce the last set of blintz folds to arrive at the finished unit.

Assembly

Assemble like Basic Blintz Units.

Finished Unit

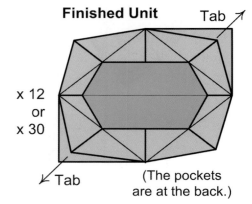

Tab

x 12 or x 30

Tab

(The pockets are at the back.)

Thirty-unit assembly of Impatient in random coloring.

Impatient Trillium Bouquet

Start with a completed and flattened Impatient unit.

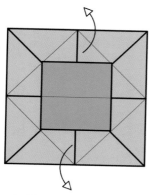

1. Unfold all the way the two flaps shown.

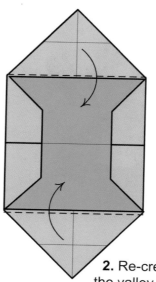

2. Re-crease the valley folds.

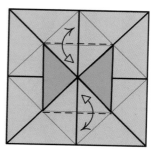

3. Valley fold and unfold along pre-existing creases.

 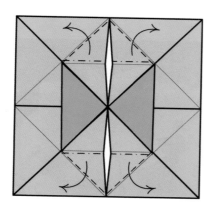

4. Squash both corners.

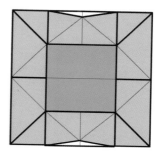

5. Unflatten back to the original Impatient unit, by reinforcing the blintz folds and the horizontal fold at the center.

Finished Unit

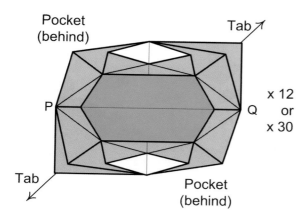

Pocket (behind)

Tab

P

Tab

Q

x 12 or x 30

Pocket (behind)

Assembly

Assemble like the Basic Blintz model. Remember that for the octahedral or icosahedral assembly, PQ will lie along the edges of the polyhedron. After assembly the squashed petals will open up automatically.

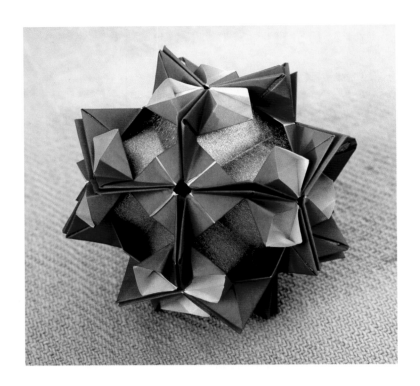

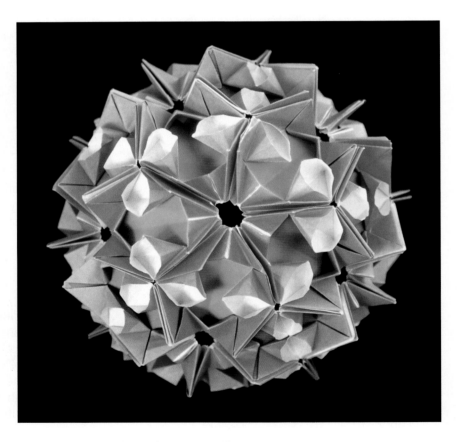

Twelve-unit and 30-unit assemblies of Impatient Trillium Bouquets.

Blintz Base Bouquets

Impatient Gypsophila Bouquet

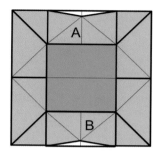

1. Start with Step 5 of the previous model, the Impatient Trillium Bouquet.

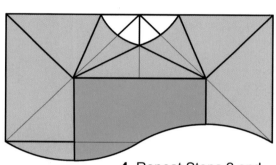

2. Valley fold and unfold. (Make soft creases.)

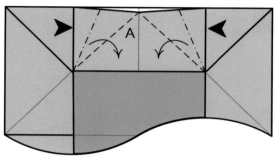

3. Squash the two sides with soft creases.

4. Repeat Steps 2 and 3 on the bottom flap B.

Finished Unit

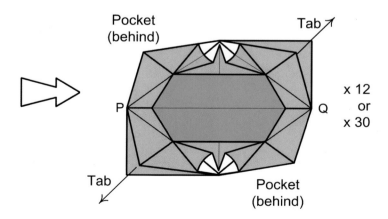

Pocket (behind)

Tab ↗

P

Q

Tab ↙

Pocket (behind)

x 12
or
x 30

Assembly

Assemble like the Basic Blintz model. Remember that for the octahedral or icosahedral assembly, PQ will lie along the edges of the polyhedron.

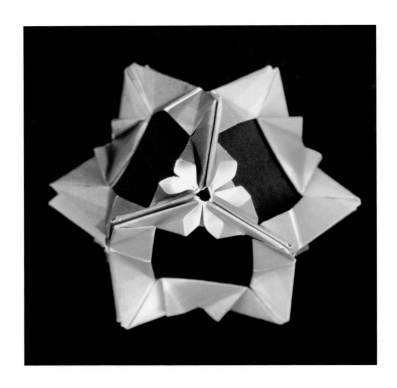

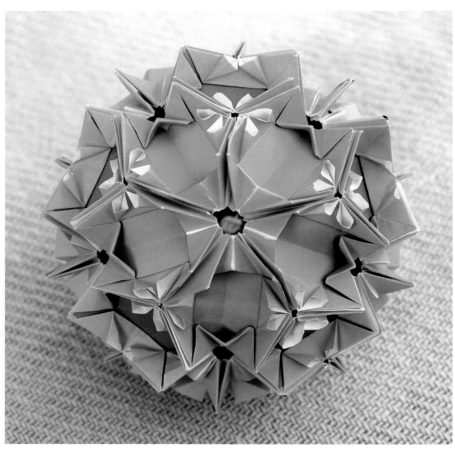

Six-unit and 30-unit assemblies of Impatient Gypsophila Bouquets.

Blintz Base Bouquets

Chrysanthemum and Combinations

Start with a completed and flattened Basic Blintz Unit with Step 10 skipped.

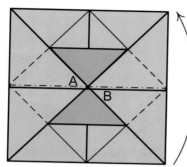

1. Bring bottom edge to top while making the mountain and valley creases on flaps A and B and bringing points A and B to the top.

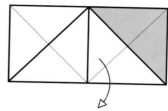

2. Unfold last step gently to arrive at the finished unit.

Assemble like Basic Blintz Units.

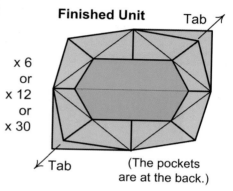

Finished Unit

Tab

x 6
or
x 12
or
x 30

Tab

(The pockets are at the back.)

Shown below are Chrysanthemum Trillium and Chrysanthemum Gypsophyla units. Please refer to respective Impatient Bouquet sections for how to make these units.

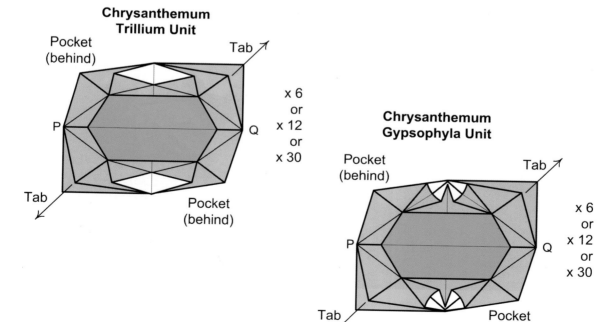

Chrysanthemum Trillium Unit

Pocket (behind)

Tab

P   Q

Tab

Pocket (behind)

x 6
or
x 12
or
x 30

Chrysanthemum Gypsophyla Unit

Pocket (behind)

Tab

P   Q

Tab

Pocket (behind)

x 6
or
x 12
or
x 30

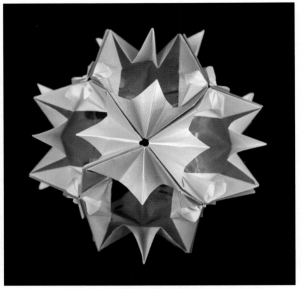

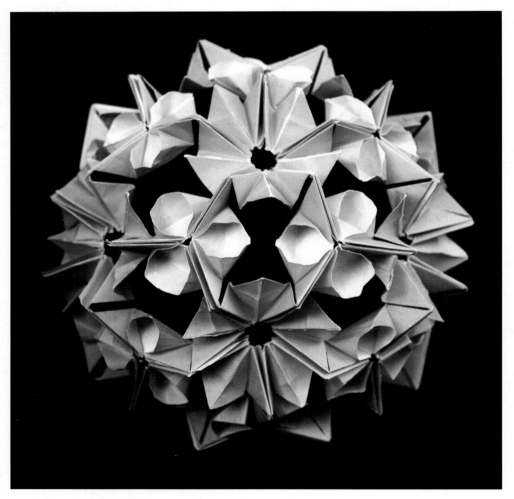

Six-unit Chrysanthemum (top left), 12-unit Chrysanthemum Gypsophila (top right), and 30-unit Chrysanthemum Trillium (bottom).

Periwinkle and Combinations

Start with a completed and flattened Basic Blintz Unit with Step 10 skipped.

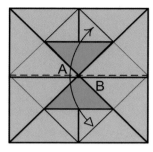

1. Valley fold through all layers and unfold.

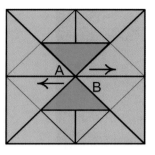

2. Curl tips A and B in the arrow directions.

Gently reinforce the last set of blintz folds to arrive at the finished unit.

Finished Unit

Assemble exactly like Basic Blintz Units.

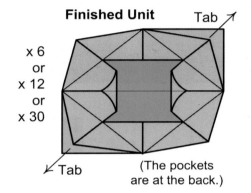

x 6
or
x 12
or
x 30

Tab

Tab

(The pockets are at the back.)

Shown below are Periwinkle Trillium and Periwinkle Gypsophyla units. Please refer to respective Impatient Bouquet sections for how to make these units.

Periwinkle Trillium Unit

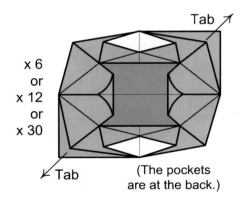

x 6
or
x 12
or
x 30

Tab

Tab

(The pockets are at the back.)

Periwinkle Gypsophyla Unit

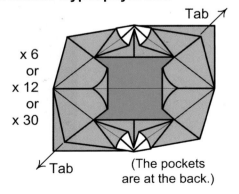

x 6
or
x 12
or
x 30

Tab

Tab

(The pockets are at the back.)

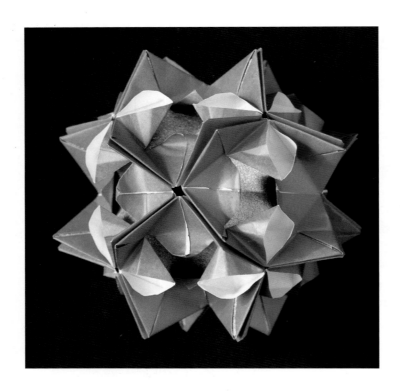

12-unit assembly of Priwinkle with petals uncurled (top) and 30-unit assembly of Periwinkle Gypsophila (bottom).

Blintz Base Models without Inserts

The units get thick, so only *kami* or thinner paper is recommended. Use at least 4" squares.

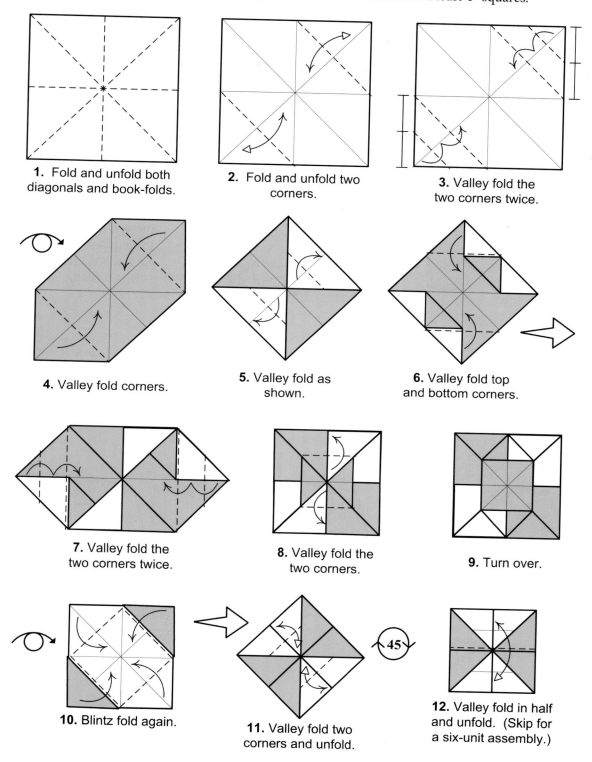

1. Fold and unfold both diagonals and book-folds.

2. Fold and unfold two corners.

3. Valley fold the two corners twice.

4. Valley fold corners.

5. Valley fold as shown.

6. Valley fold top and bottom corners.

7. Valley fold the two corners twice.

8. Valley fold the two corners.

9. Turn over.

10. Blintz fold again.

11. Valley fold two corners and unfold.

12. Valley fold in half and unfold. (Skip for a six-unit assembly.)

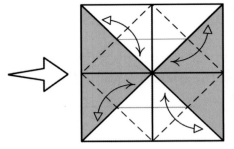

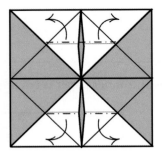

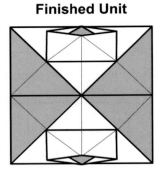

13. Blintz fold firmly through all layers and unfold.

14. Squash fold the two corners shown.

Front

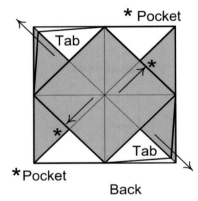

* Pocket

Tab

*

*

Tab

*Pocket

Back

x 6 or x 12 or x 30

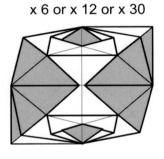

Reinforce the blintz folds from Step 13. The finished unit becomes three dimensional and the squash folds open up a bit. Assemble like Basic Blintz Units.

Note that the unit shown above is a Periwinkle/Trillium unit with colored Periwinkles and white Trillium (colored insides). Try other flower combinations presented earlier in the chapter with this non-insert version. You may play with the coloring as well. Try these other colorings illustrated below by simply reversing some folds from valleys to mountains.

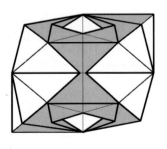

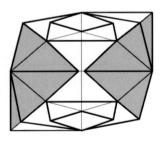

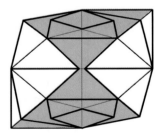

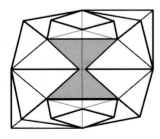

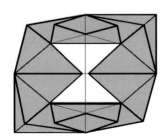

Impatient Gypsophila in duo paper (top) and Chrysanthemum Trillium without inserts.

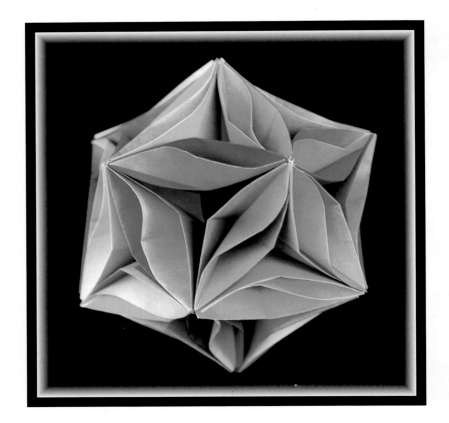

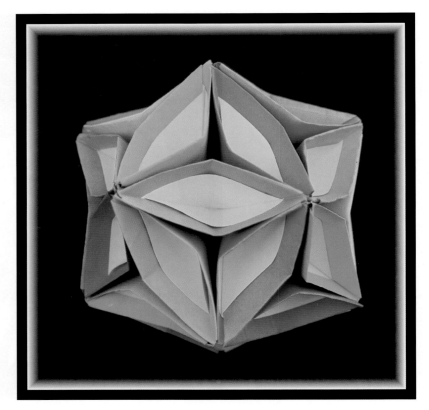

Icosahedron with Curves 2 (top) and Icosahedron with Waves (bottom).

4 ◆ Decorative Icosahedra

In this chapter, five decorative icosahedra are presented. The first four models all belong to a family that I have named Icosahedra with Curves. The units are very simple and some involve inserts. Contrasting insert colors produce visually stunning results. For these models assembly aids such as clothespins are recommended for beginners.

The last icosahedron of the chapter, the Patterned Icosahedron, is based on Lewis Simon and Bennett Arnstein's Triangle Edge Module from the book *3-D Geometric Origami: Modular Polyhedra* by Rona Gurkewitz and Bennett Arnstein. One pattern variation has been presented but many other variations are possible. I give much thanks to the above mentioned people for allowing me to use their basic unit from which my patterns are derived.

All of the models in this chapter use half squares, i.e., 2:1 rectangles. The insert sizes for the Curved Icosahedra vary but those sizes in turn are derived from the paper for the main unit. For all of the Icosahedra with Curves models, color photocopy paper works very well because the back side of the paper is not exposed and the thickness is perfect. However, *kami* or virtually any other paper can be used. For the Patterned Icosahedron, *kami* is a must because the back side needs to be exposed to create the motifs.

A note on assembly: It is important that the units meet at the vertices during assembly. This results in the center of the model opening up a bit to generate a hollow space, which is perfectly fine.

Recommendations

Paper Size: 2″–3″ wide rectangles

Paper Type: Photocopy or any other paper

Finished Model Size

2.5″ wide rectangles yield 30-unit

models that are 3.5″ in height approximately

Icosahedron with Curves

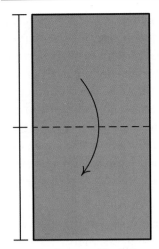

1. Start with 1:2 paper and fold in half.

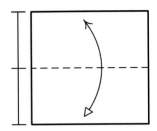

2. Valley fold and unfold top layer only.

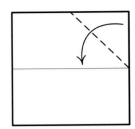

3. Valley fold corner.

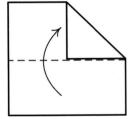

4. Re-crease fold from Step 2.

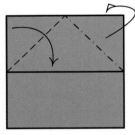

5. Valley fold left corner, top layer only. Mountain fold right corner so flap goes behind all layers.

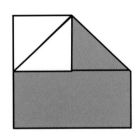

6. Turn over.

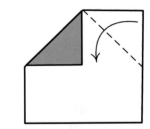

7. Valley fold corner.

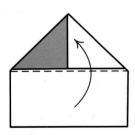

8. Valley fold.

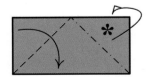

9. Valley fold left corner. Mountain fold right corner and tuck flap in the slit behind the fold.

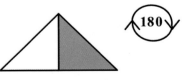

10. Turn around.

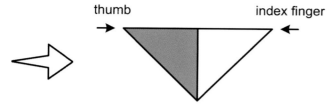

thumb   index finger

Finished Unit

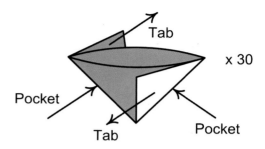

Tab

x 30

Pocket

Tab   Pocket

11. Gently apply pressure to the corners with your thumb and index finger to open up the top. Then curve the top edges outwards with your nails.

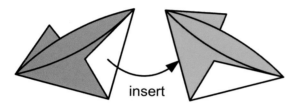

insert

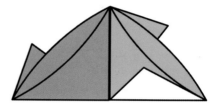

Tetrahedral Assembly

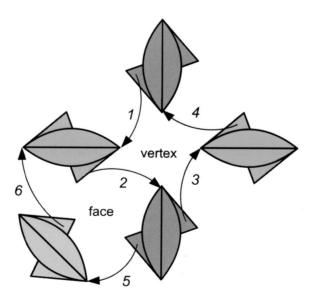

vertex

face

1 2 3 4 5 6

To make a tetrahedron, assemble four units in a ring to form a vertex following the sequence numbers. Insert a fifth unit to form a face by performing Steps 5 and 6. Continue assembling in this manner to form six vertices and eight faces to arrive at the finished model. Assembly aids such as miniature clothespins may be used and they can be removed as assembly progresses or at the end.

Icosahedral Assembly

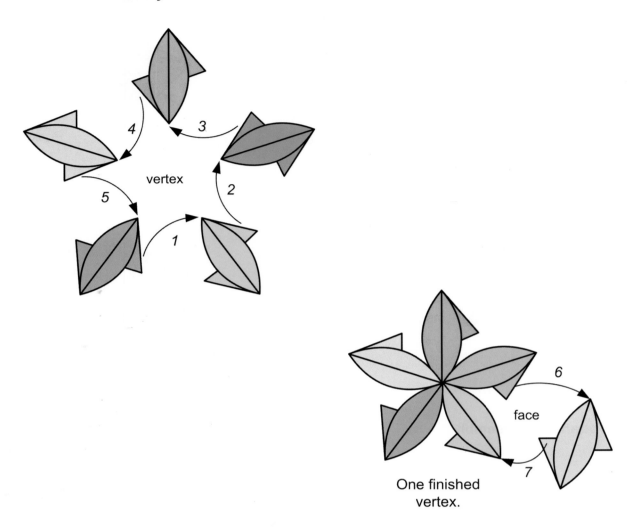

vertex

One finished
vertex.

face

To make an icosahedron, assemble five units in a ring to form a vertex following the
sequence numbers. Insert a sixth unit to form a face by performing Steps 6 and 7.
Continue assembling in this manner to form 12 vertices and 20 faces to arrive at the
finished model. Assembly aids such as miniature clothespins may be used and they
can be removed as assembly progresses or at the end.

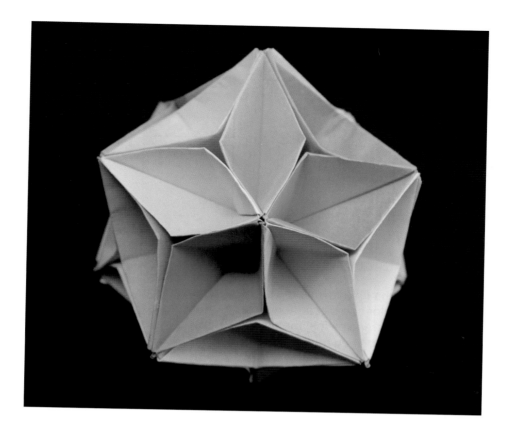

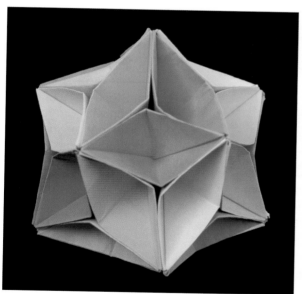

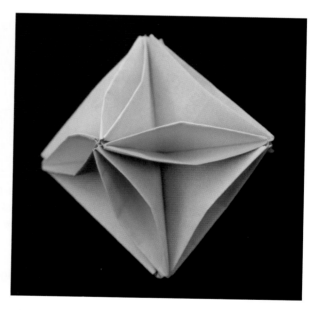

Two views of Icosahedron with Curves (top and bottom left) made in a five-color scheme and a tetrahedral assembly in a three-color scheme (bottom right).

Icosahedron with Curves 2

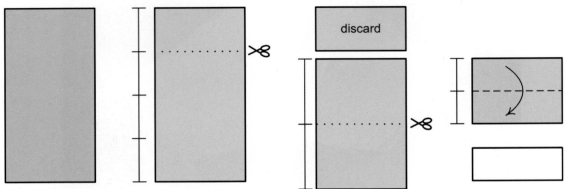

1:2 paper for _main unit_.

For the *inserts*, use the same size paper as the main unit of a contrasting color and discard one quarter of the length. Cut the remainder into halves. Valley fold the new rectangle into half to complete an insert. Note that you may try other insert widths as well for a different look.

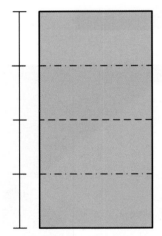

1. Start with main unit paper, fold as shown and unfold.

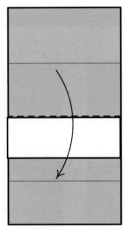

2. Place insert as shown with open ends downwards and re-crease valley fold.

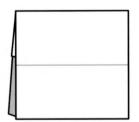

3. Repeat Steps 2–10 of the Icosahedron with Curves, making sure that the insert does not move from its initial position and stays put.

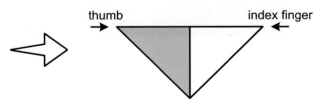

Gently apply pressure to the corners with your thumb and index finger to open up the top as well as the insert. Then firmly curve the top edges outwards with your nails.

thumb   index finger

Finished Unit

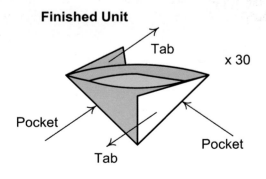

Tab

x 30

Pocket

Pocket

Tab

Assembly

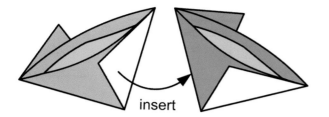

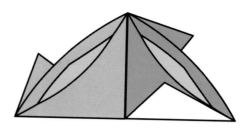

Insert tab into pocket. Assemble to
make an icosahedron as explained in
the Icosahedron with Curves model.

One finished
vertex.

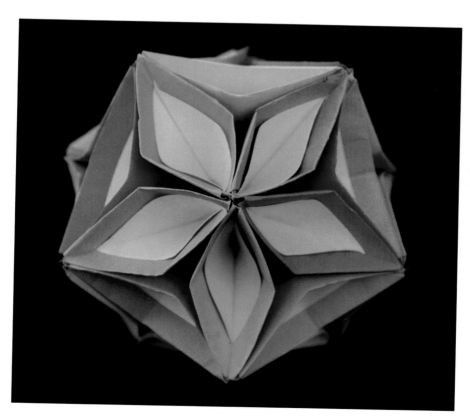

Icosahedron with Curves 2.

Icosahedron with Curves and Waves

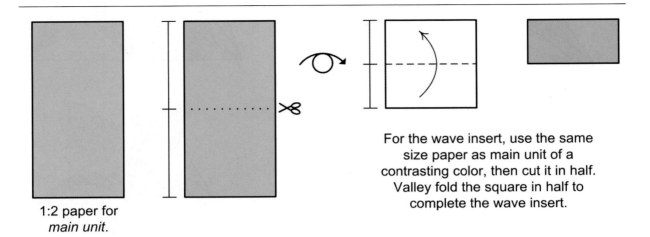

For the wave insert, use the same size paper as main unit of a contrasting color, then cut it in half. Valley fold the square in half to complete the wave insert.

1:2 paper for *main unit*.

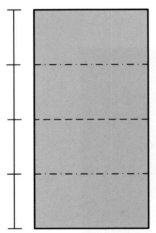

1. Start with main unit paper, fold as shown and unfold.

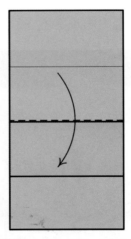

2. Place insert as shown with open ends upwards and re-crease valley fold.

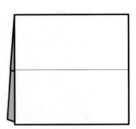

3. Repeat Steps 2–10 of the Icosahedron With Curves, making sure that the insert does not move from its initial position and stays put.

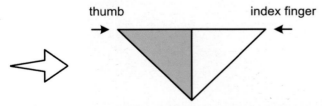

thumb   index finger

Gently apply pressure to the corners with your thumb and index finger to open up the top. Then firmly curve the top edges outwards with your nails. The insert will automatically take the shape of a sine wave. Gently reinforce it.

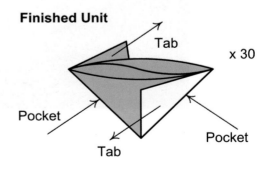

Finished Unit

Tab   x 30

Pocket

Tab

Pocket

Assembly

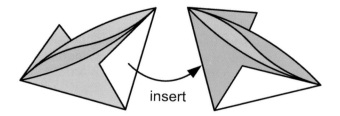

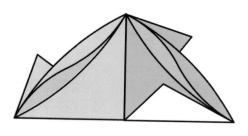

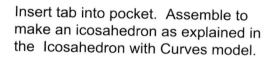

insert

Insert tab into pocket. Assemble to
make an icosahedron as explained in
the Icosahedron with Curves model.

One finished
vertex.

Icosahedron with Curves and Waves.

Icosahedron with Curves and Petals

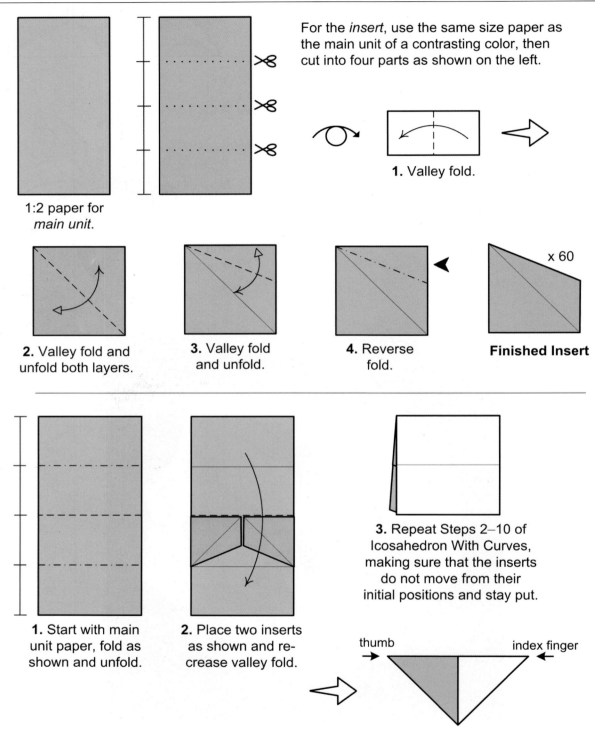

For the *insert*, use the same size paper as the main unit of a contrasting color, then cut into four parts as shown on the left.

1. Valley fold.

1:2 paper for *main unit*.

2. Valley fold and unfold both layers.

3. Valley fold and unfold.

4. Reverse fold.

x 60

Finished Insert

1. Start with main unit paper, fold as shown and unfold.

2. Place two inserts as shown and re-crease valley fold.

3. Repeat Steps 2–10 of Icosahedron With Curves, making sure that the inserts do not move from their initial positions and stay put.

thumb   index finger

4. Gently apply pressure to the corners with your thumb and index finger to open up the top. Then firmly curve the top edges outwards with your nails. Open up inserts as well and curve them like petals.

Finished Unit

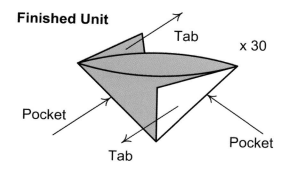

Tab

x 30

Pocket

Tab

Pocket

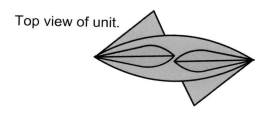

Top view of unit.

Assembly

Insert tab into pocket. Assemble to make an icosahedron as explained in the Icosahedron with Curves model.

One finished vertex with white petals.

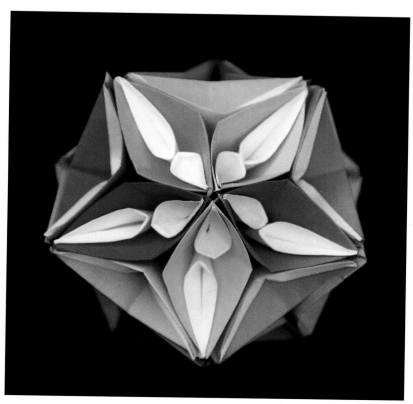

Icosahedron with Curves and Petals.

Patterned Icosahedron

The patterned icosahedron presented here is an adaptation of a unit called the Triangle Edge Module by Lewis Simon and Bennett Arnstein (Gurkewitz & Arnstein, 1995). This adaptation merely exposes the white side of the paper to create interesting patterns on the finished faces and vertices of the icosahedron.

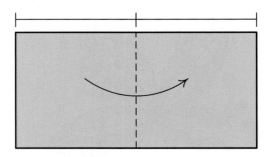

1. Start with 1:2 paper and make a book-fold.

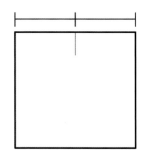

2. Pinch midpoint of top edge of top layer only.

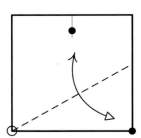

3. Valley fold to match dots and unfold.

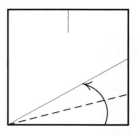

4. Valley fold bottom edge to new crease.

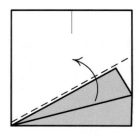

5. Re-crease fold from Step 3.

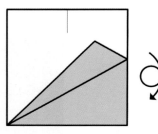

6. Turn over *vertically*.

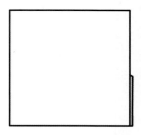

7. Repeat Steps 2–5 on this side.

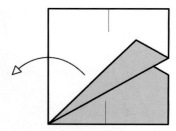

8. Unfold toward the left.

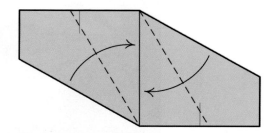

9. Valley fold edges to center.

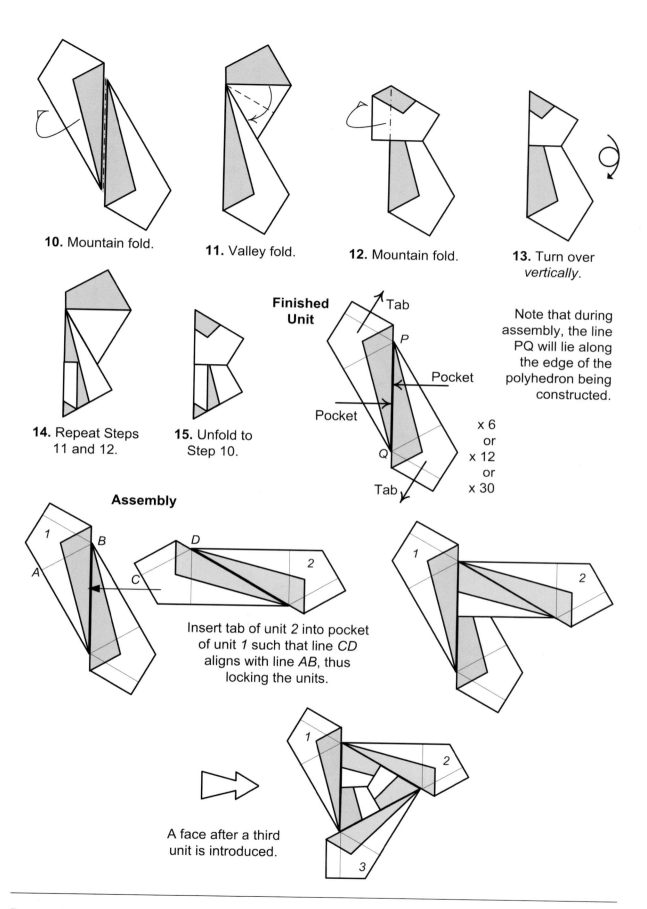

10. Mountain fold.

11. Valley fold.

12. Mountain fold.

13. Turn over *vertically*.

Note that during assembly, the line PQ will lie along the edge of the polyhedron being constructed.

14. Repeat Steps 11 and 12.

15. Unfold to Step 10.

Finished Unit

Tab

P

Pocket

Pocket

Q

Tab

x 6
or
x 12
or
x 30

Assembly

1

B

A

D

C

2

Insert tab of unit *2* into pocket of unit *1* such that line *CD* aligns with line *AB*, thus locking the units.

1

2

A face after a third unit is introduced.

1

2

3

Six-Unit
Tetrahedral Assembly

Connect three units in a ring to form a vertex. Add another unit to form a triangular face. Continue assembling to form four vertices and four faces to complete a tetrahedron.

Twelve-Unit
Octahedral Assembly

Connect four units in a ring to form a vertex. Add another unit to form a triangular face. Continue assembling to form six vertices and eight faces to complete an octahedron.

Thirty-Unit
Icosahedral Assembly

Connect five units in a ring to form a vertex. Add another unit to form a triangular face. Continue assembling to form 12 vertices and 20 faces to complete an icosahedron.

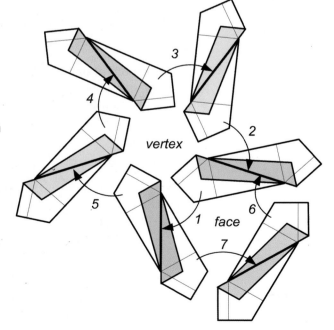

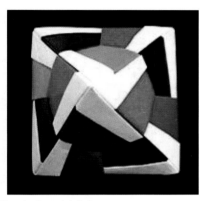

Thirty-unit Patterned Icosahedron (left) and 12-unit Patterned Octahedron (right)
with regular and reverse coloring, respectively.

Two other possible pattern variations shown in six-unit tetrahedral (left) and 30-unit icosahedral (right) assemblies. These
variations are minor and not diagrammed.

Thirty-unit assemblies of Flowered Sonobe (top) and Geranium (bottom).

5 ◆ Embellished Sonobes

We will explore Sonobe-type units in this chapter. The original Sonobe Unit was created by Mitsunobu Sonobe in the late 1960s and has remained one of the foundations of polyhedral modular origami ever since. There have been many Sonobe variations and enhancements but they all lock together the same way and have similar properties. The following is an illustration of how Sonobe units lock together and how 12 or 30 units are assembled based on an underlying octahedron or icosahedron, respectively. Other assemblies based on different underlying polyhedra configurations are possible as well but are not shown here. Matte foil paper works best for the models that involve curling, while harmony paper is effective for the others. Units made with rectangles 4″ wide yield 30-unit models of approximately 4.75″ in height.

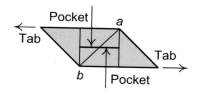

Diagrammatic representation of a generic Sonobe unit. (Note that the pockets may appear different in Sonobe variations.)

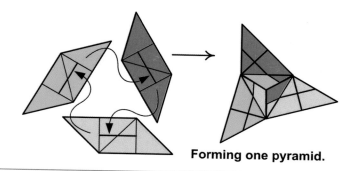

Forming one pyramid.

Assembly of 12 Sonobe-Type Units

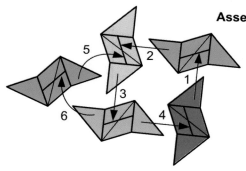

Assemble four units in a ring as shown following the sequence numbers. Take a fifth unit and perform Steps 5 and 6 to form a pyramid. Continue adding three more units to form a ring of four pyramids. Complete model by forming a total of eight pyramids arranged in an octahedral symmetry. The diagonals *ab* of a unit will lie along the edges of an octahedron.

Assembly of 30 Sonobe-Type Units

Assemble five units in a ring as shown following the sequence numbers. Take a sixth unit and perform Steps 6 and 7 to form a pyramid. Continue adding four more units to form a ring of five pyramids. Complete model by forming a total of 20 pyramids arranged in an icosahedral symmetry. The diagonals *ab* of a unit will lie along the edges of an icosahedron.

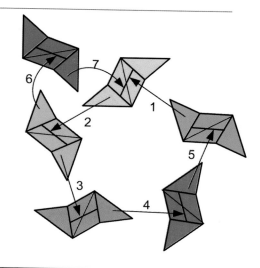

Sonobe with Fins

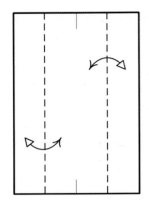

1. Start with 2:3 paper, pinch ends of book-fold, then cupboard fold and unfold.

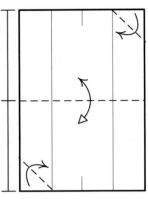

2. Fold in half and unfold, then valley fold corners.

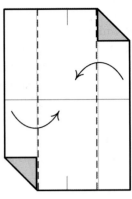

3. Re-crease cupboard folds.

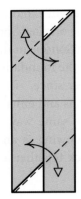

4. Valley fold and unfold.

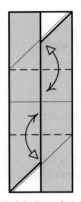

5. Valley fold and unfold.

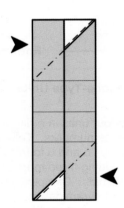

6. Inside reverse fold.

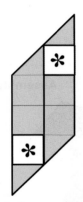

7. Tuck flaps marked ✱ in openings underneath.

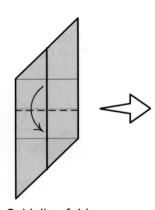

8. Valley fold along existing crease.

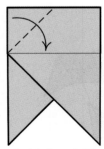

9. Valley fold corner.

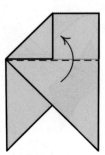

10. Valley fold top flap.

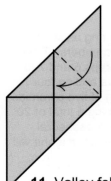

11. Valley fold.

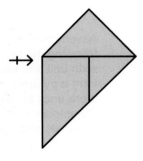

12. Repeat Steps 9–11 at the back.

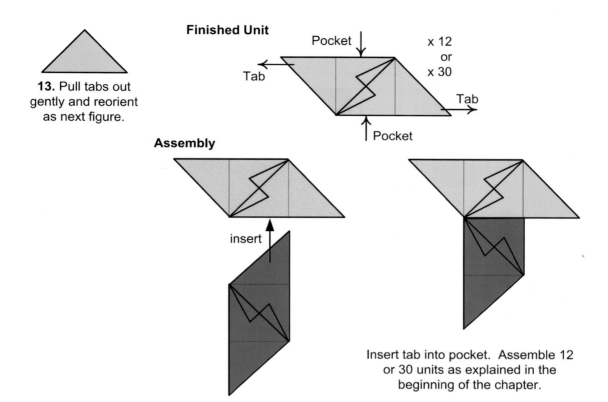

Finished Unit

Pocket → ✕ 12 or ✕ 30

Tab ←

Tab →

Pocket ↑

13. Pull tabs out gently and reorient as next figure.

Assembly

insert ↑

Insert tab into pocket. Assemble 12 or 30 units as explained in the beginning of the chapter.

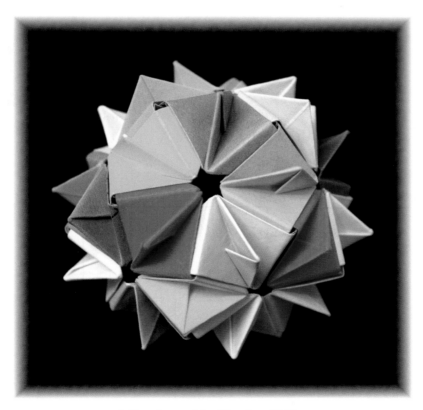

A 30-unit assembly of Sonobe with Fins.

Flowered Sonobe

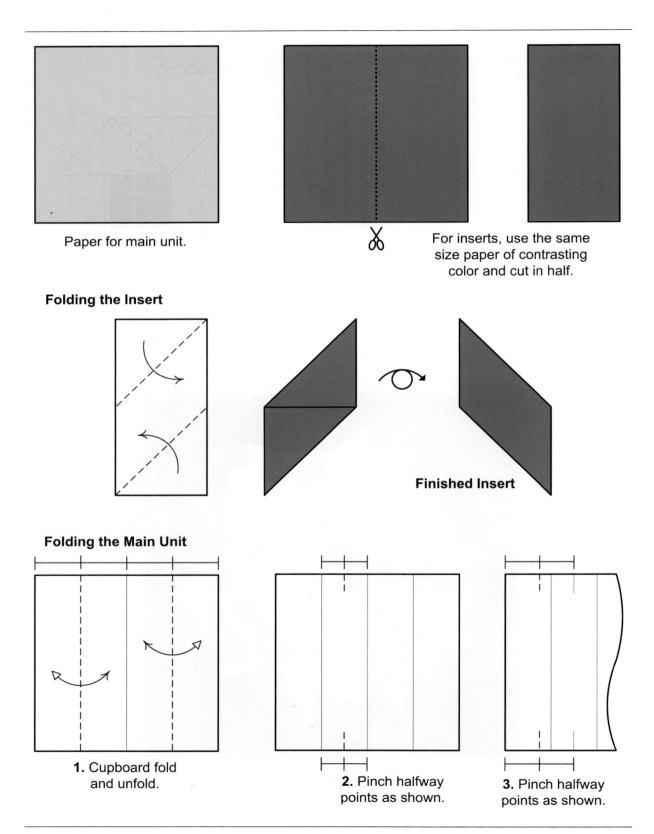

Paper for main unit.

For inserts, use the same size paper of contrasting color and cut in half.

Folding the Insert

Finished Insert

Folding the Main Unit

1. Cupboard fold and unfold.

2. Pinch halfway points as shown.

3. Pinch halfway points as shown.

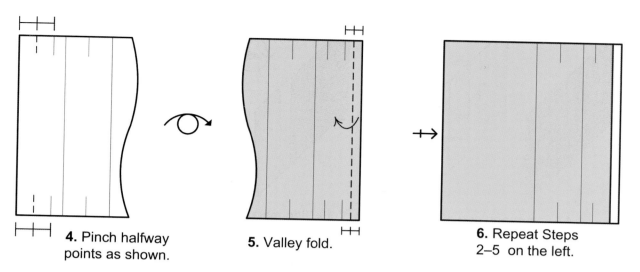

4. Pinch halfway points as shown.

5. Valley fold.

6. Repeat Steps 2–5 on the left.

Note that if you are good at folding by estimate, just fold very narrow strips on left and right edges without going through Steps 2–6.

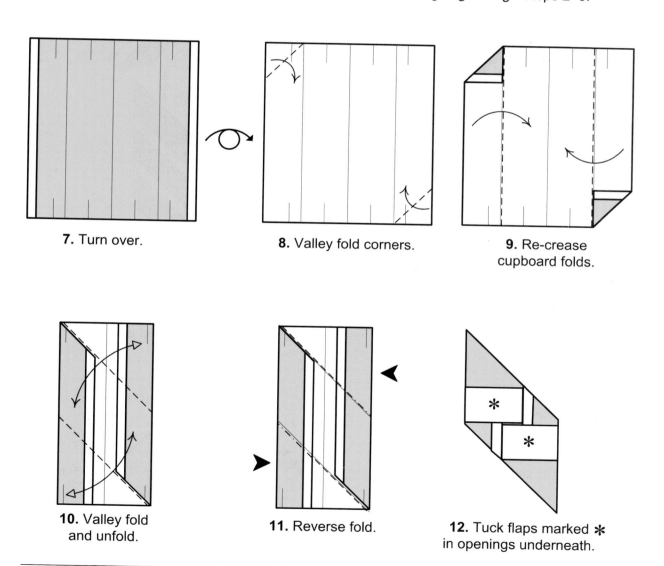

7. Turn over.

8. Valley fold corners.

9. Re-crease cupboard folds.

10. Valley fold and unfold.

11. Reverse fold.

12. Tuck flaps marked * in openings underneath.

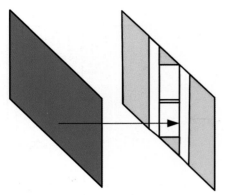

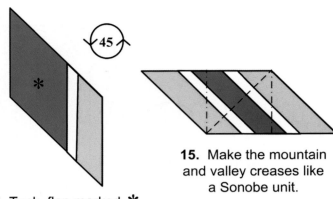

13. Insert one end of prepared insert as shown as far as it will go.

14. Tuck flap marked ✳ in opening underneath.

15. Make the mountain and valley creases like a Sonobe unit.

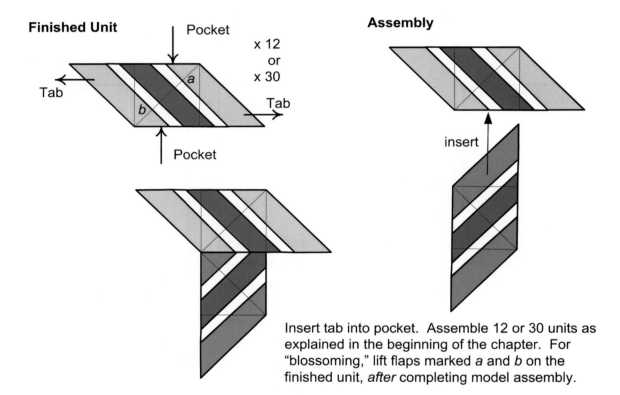

Finished Unit

Pocket

x 12 or x 30

Tab

a

b

Tab

Pocket

Assembly

insert

Insert tab into pocket. Assemble 12 or 30 units as explained in the beginning of the chapter. For "blossoming," lift flaps marked *a* and *b* on the finished unit, *after* completing model assembly.

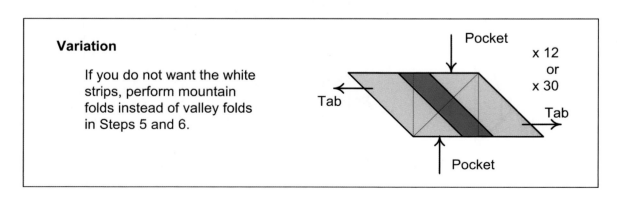

Variation

If you do not want the white strips, perform mountain folds instead of valley folds in Steps 5 and 6.

Pocket

x 12 or x 30

Tab

Tab

Pocket

Thirty-unit assemblies of Flowered Sonobe before and after blossoming (top left and right) and Flowered Sonobe Variation (bottom).

Dahlia

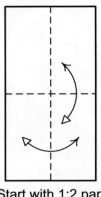

1. Start with 1:2 paper. Fold both book-folds and unfold.

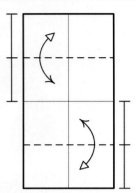

2. Valley fold and unfold.

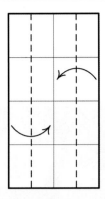

3. Cupboard fold.

4. Valley fold and unfold marked diagonals.

5. Repeat for the other two diagonals, then unfold all.

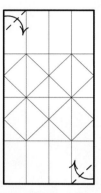

6. Valley fold corners.

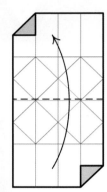

7. Valley fold pre-existing crease.

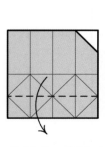

8. Valley fold as shown.

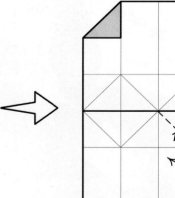

9. Valley and mountain fold following the sequence numbers. Folds *1* and *2* involve two layers.

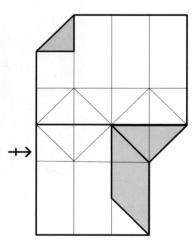

10. Repeat Step 9 on the left.

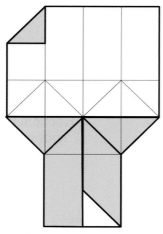

11. Turn over.

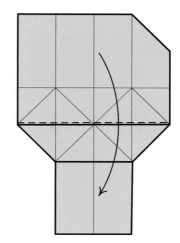

12. Valley fold.

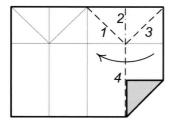

13. Repeat Step 9.

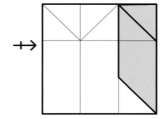

14. Repeat step 9 on the left.

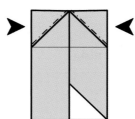

15. Reverse fold two corners. Make sure the folds go into the *center* of all layers.

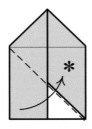

16. Valley fold and tuck under the flap marked ✳.

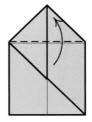

17. Valley fold.

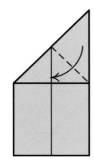

18. Valley fold.

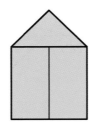

19. Turn over.

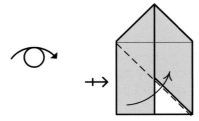

20. Repeat Steps 16–18.

21. Pull down flaps both in the front and back.

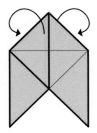

22. Curl front tip toward you and back tip away from you with a narrow pencil-like object.

Reorient unit as shown below.

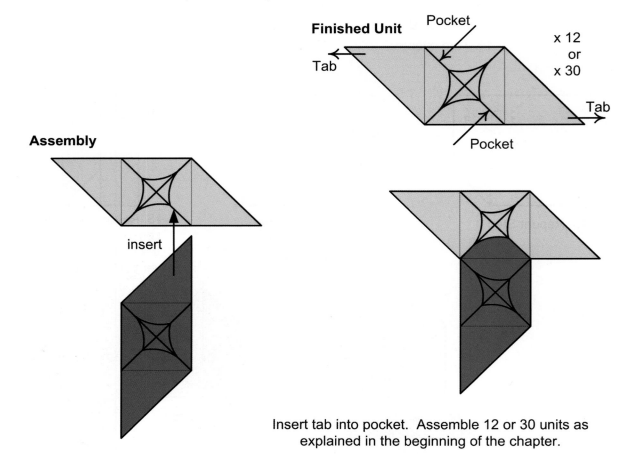

Finished Unit

Pocket

Tab

x 12 or x 30

Tab

Pocket

Assembly

insert

Insert tab into pocket. Assemble 12 or 30 units as explained in the beginning of the chapter.

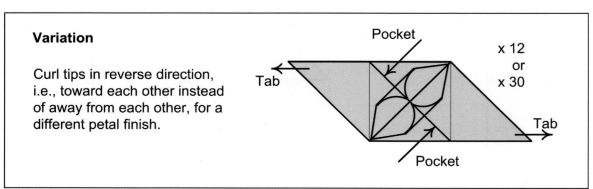

Variation

Curl tips in reverse direction, i.e., toward each other instead of away from each other, for a different petal finish.

Pocket

Tab

x 12 or x 30

Tab

Pocket

Dahlias with petals curled outwards (top) and inwards (bottom).

Geranium

This model is similar to the Dahlia model but has longer and pointier petal tips. Start with any rectangle of proportion between 1:2.5 and 1:(1+√3), i.e., between 1:2.5 and 1:2.73. The diagrams below use 1:(1+√3) paper.

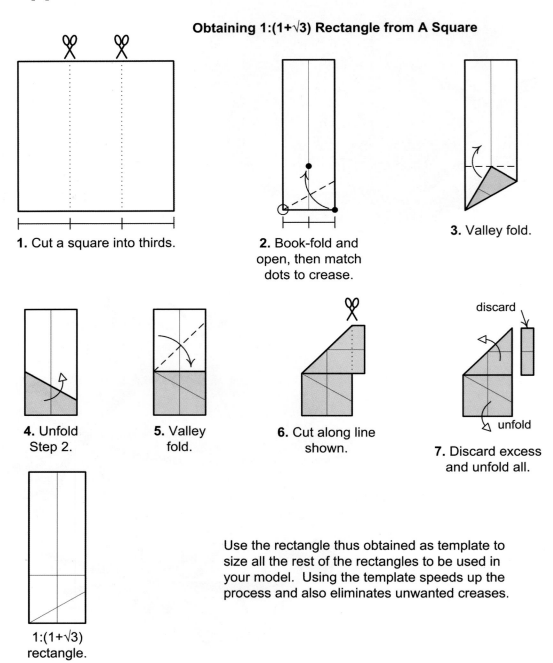

Obtaining 1:(1+√3) Rectangle from A Square

1. Cut a square into thirds.

2. Book-fold and open, then match dots to crease.

3. Valley fold.

4. Unfold Step 2.

5. Valley fold.

6. Cut along line shown.

discard

unfold

7. Discard excess and unfold all.

1:(1+√3) rectangle.

Use the rectangle thus obtained as template to size all the rest of the rectangles to be used in your model. Using the template speeds up the process and also eliminates unwanted creases.

After sizing your rectangles as explained in the previous page, fold as follows to make the Geranium units.

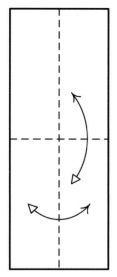 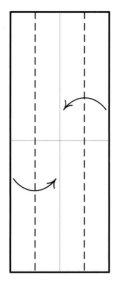  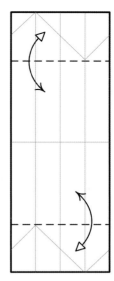

1. Crease both book-folds and unfold.

2. Cupboard fold.

3. Valley fold corners, then unfold all.

4. Valley fold and unfold as shown, then re-crease cupboard folds.

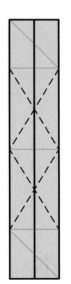 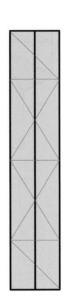 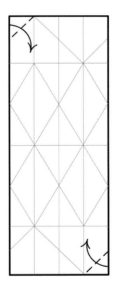 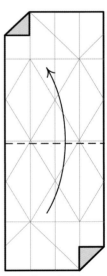

5. Valley fold and unfold marked diagonals.

6. Unfold to Step 1.

7. Valley fold corners.

8. Valley fold pre-existing crease.

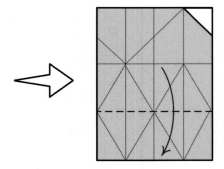

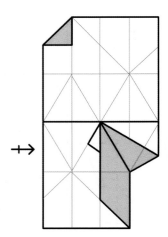

9. Valley fold as shown.

10. Valley and mountain fold following sequence numbers. (Note that folds *1* and *2* involve two layers.)

11. Repeat Step 9 on the left.

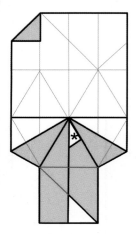

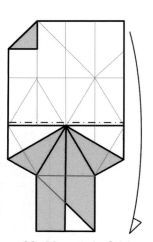

12. Tuck white corner marked ✷ in flap underneath.

13. Mountain fold and turn over.

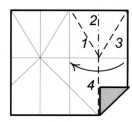

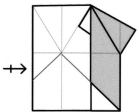

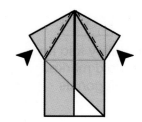

14. Repeat Step 10.

15. Repeat Steps 11 and 12 on the left.

16. Reverse fold two corners. Make sure the folds go into the *center* of all layers.

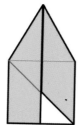

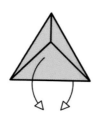

18. Pull down flaps both in the front and back.

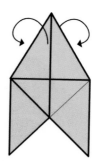

19. Curl front tip toward you and back tip away from you with a narrow pencil-like object.

17. Perform steps 16–20 of the previous model, the Dahlia.

Reorient unit as shown on the right. Insert tab into pocket. Assemble 12 or 30 units as explained in the beginning of the chapter.

Finished Unit

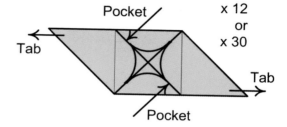

Pocket

Tab

Pocket

Tab

x 12 or x 30

Assembly

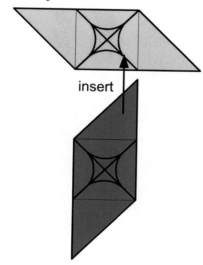

insert

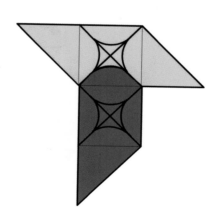

Variation

Curl tips in reverse direction, i.e., towards each other instead of away from each other, for a different petal finish.

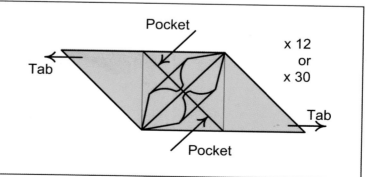

Pocket

Tab

Pocket

Tab

x 12 or x 30

A 12-unit Geranium in regular curls (top) and a 30-unit Geranium in reverse curls (bottom).

Embellished Sonobes

Zinnia

Harmony paper is recommended for Zinnia for contrast between petals and background.

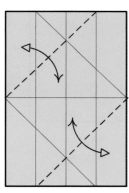

1. Start with 2:3 paper and crease both book-folds.

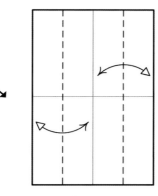

2. Cupboard fold and unfold.

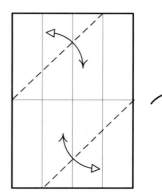

3. Valley fold and unfold.

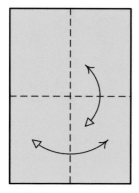

4. Valley fold and unfold.

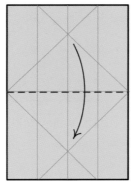

5. Re-crease valley fold.

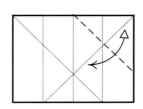

6. Valley fold corner, then unfold all.

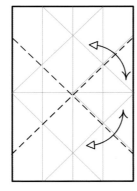

7. Fold and unfold the diagonals shown, using creases made in Step 6.

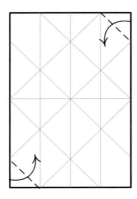

8. Valley fold corners shown.

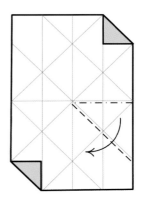

9. Valley and mountain fold pre-existing creases.

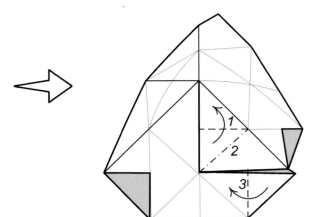

10. Valley and mountain fold following the sequence numbers. Steps *1* and *2* involve two layers of paper.

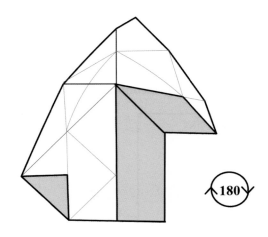

11. Turn around.

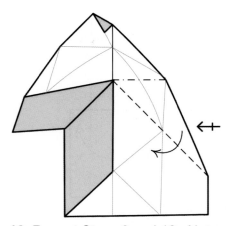

12. Repeat Steps 9 and 10. Note that it gets a bit difficult and you need to reach into invisible areas.

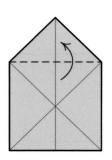

13. Valley fold.

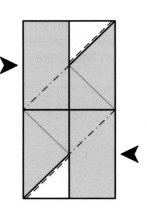

14. Inside reverse fold.

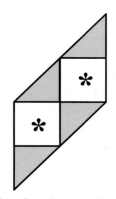

15. Gently tuck flaps marked ✳ in openings underneath, taking care not to rip paper.

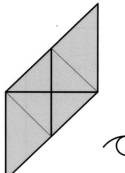

16. Turn over.

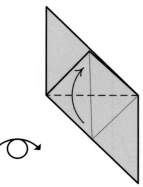

17. Valley fold.

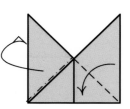

18. Valley and mountain fold.

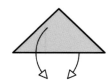

19. Pull down flaps both in the front and back.

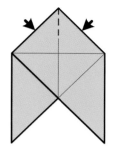

20. Gently spread apart openings marked by arrows and give them a curved shape. Valley fold tip partway.

21. Reorient like the next figure.

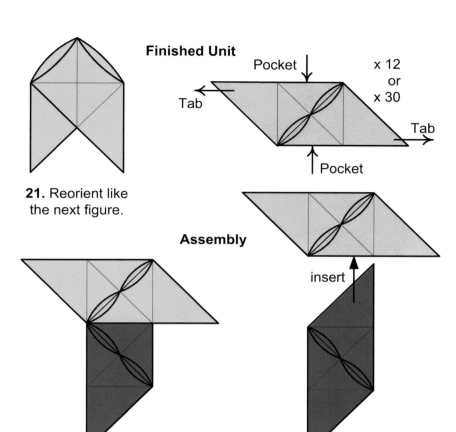

Finished Unit

Pocket

Tab

x 12
or
x 30

Tab

Pocket

Assembly

insert

Insert tab into pocket. Assemble 12 or 30 units as explained in the beginning of the chapter.

A 30-unit assembly of Zinnia made with harmony paper.

Layered Petunia (top) and Daisy (bottom).

Embellished Floral Balls

6 ◈ Embellished Floral Balls

The models presented in this chapter have the common property of locking together in a similar fashion. These models are embellishments of the original Floral Balls in my first book *Marvelous Modular Origami* [Muk07]. The models are each made of 30 units and they have an icosahedral/dodecahedral symmetry. Because the units get thick and curling is involved in most, matte foil origami paper is recommended although *kami* also works. Foil-backed paper has the additional advantage of producing firmer locks. Please note that like many other modular models, the assembly will appear spread out and unruly until the last flower is in place.

For all of the models except the first two, where the back side of the paper is significantly exposed, some simple inserts can be employed to easily simulate the effect of duo paper. Simulated duo paper is often useful because duo paper is not readily available, especially in packages of a single pair of colors. The use of the inserts is not being diagrammed here and is left up to the reader to experiment with them. The length of the inserts can be of about a quarter to half of the length of the original rectangle. The width of the inserts should remain the same as that of the original rectangles.

Explained below is how a model is assembled. Please use this guide to assemble any model in this chapter. It is possible to make hexagonal ornaments as well with the units. These six-unit ornaments are discussed at the end of this chapter on page 103.

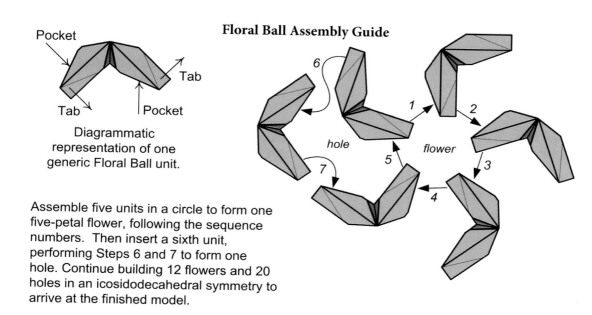

Floral Ball Assembly Guide

Diagrammatic representation of one generic Floral Ball unit.

Assemble five units in a circle to form one five-petal flower, following the sequence numbers. Then insert a sixth unit, performing Steps 6 and 7 to form one hole. Continue building 12 flowers and 20 holes in an icosidodecahedral symmetry to arrive at the finished model.

Recommendations

Paper Size: 2″–3″ wide rectangles— lengths vary with model

Paper Type: Matte foil paper or *kami*

Finished Model Size

3″ wide rectangles yield models that are approximately 5.5″ in height

Daylily

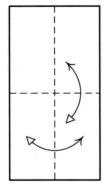

1. Start with 1:2 paper. Crease both book-folds and unfold.

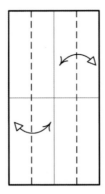

2. Cupboard fold and unfold.

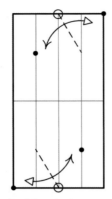

3. Match dots to form new creases.

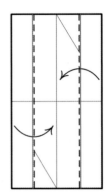

4. Re-crease cupboard folds.

5. Valley fold and unfold.

6. Valley fold, then unfold all.

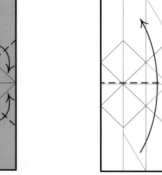

7. Valley fold pre-existing crease.

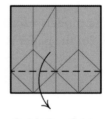

8. Valley fold as shown.

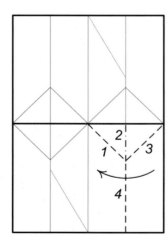

9. Valley and mountain fold following the sequence numbers. Folds *1* and *2* involve two layers.

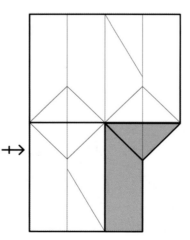

10. Repeat Step 9 on the left.

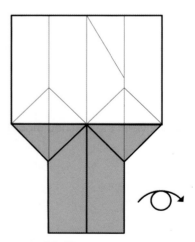

11. Turn over.

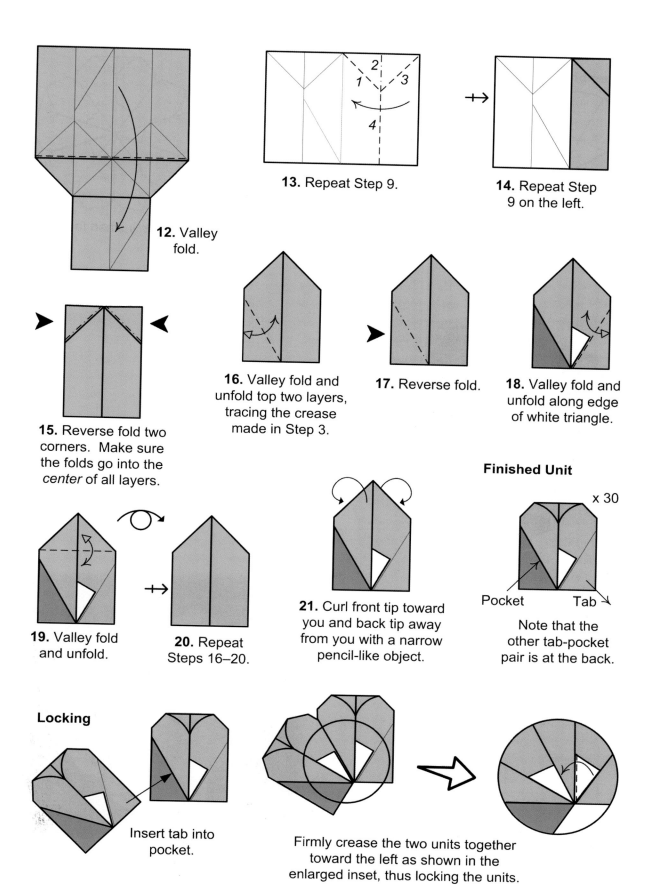

12. Valley fold.

13. Repeat Step 9.

14. Repeat Step 9 on the left.

15. Reverse fold two corners. Make sure the folds go into the *center* of all layers.

16. Valley fold and unfold top two layers, tracing the crease made in Step 3.

17. Reverse fold.

18. Valley fold and unfold along edge of white triangle.

19. Valley fold and unfold.

20. Repeat Steps 16–20.

21. Curl front tip toward you and back tip away from you with a narrow pencil-like object.

Finished Unit

x 30

Pocket   Tab

Note that the other tab-pocket pair is at the back.

Locking

Insert tab into pocket.

Firmly crease the two units together toward the left as shown in the enlarged inset, thus locking the units.

Assembly

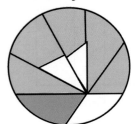

Two units
locked.

Assemble 30 units in an
icosidodecahedral symmetry
as explained in the beginning
of the chapter. Assembly gets
difficult toward the end and
tweezers may be required.

(Note that you may also try
the hidden lock method of the
next model, the Waterlily.)

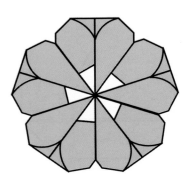

One finished face.

Two views of Daylily made with duo paper (top) and *kami* (bottom).

Embellished Floral Balls

Waterlily

Start with 1:3 rectangles. (Please see *Origami Symbols and Bases* in Chapter 1 to crease a square into thirds and thus obtain 1:3 rectangles.) The folds and assembly are similar to Daylily but the angles of the petal tips are more acute.

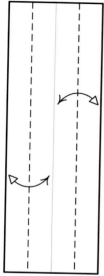

1. Crease book-fold, then cupboard fold and unfold.

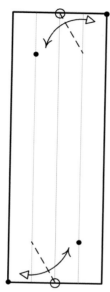

2. Match dots to form new creases.

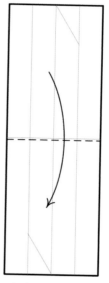

3. Valley fold into half.

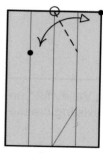

4. Crease to match dots as shown.

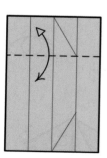

5. Valley fold both layers, then unfold all.

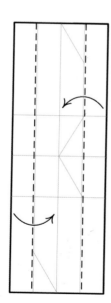

6. Re-crease cupboard folds.

7. Valley fold and unfold marked diagonals.

8. Unfold all.

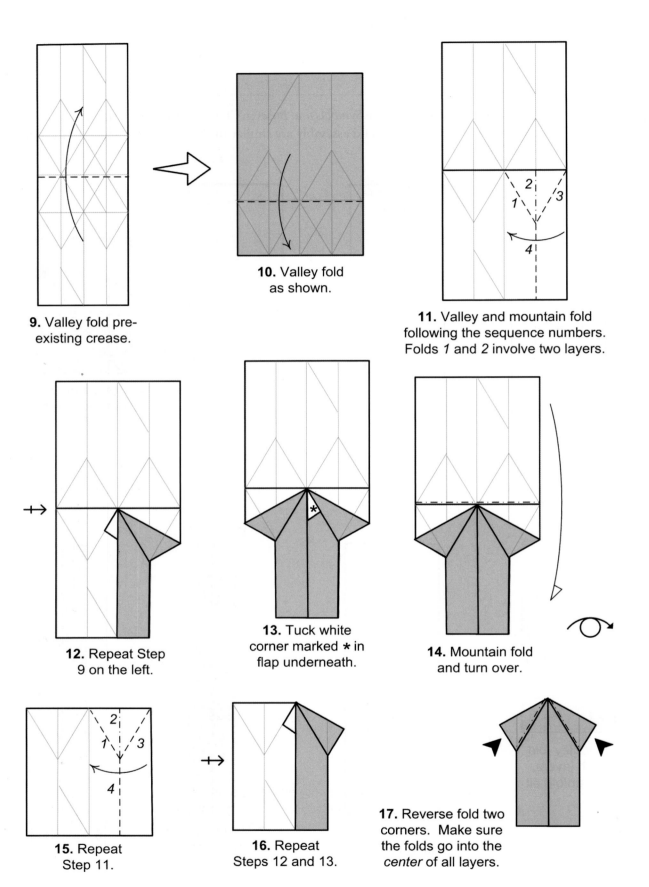

9. Valley fold pre-existing crease.

10. Valley fold as shown.

11. Valley and mountain fold following the sequence numbers. Folds *1* and *2* involve two layers.

12. Repeat Step 9 on the left.

13. Tuck white corner marked ✳ in flap underneath.

14. Mountain fold and turn over.

15. Repeat Step 11.

16. Repeat Steps 12 and 13.

17. Reverse fold two corners. Make sure the folds go into the *center* of all layers.

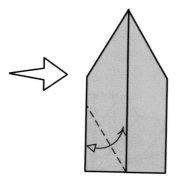
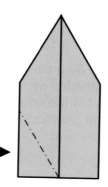
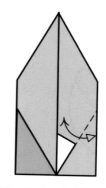

18. Valley fold and unfold top two layers following the crease made in Step 2.

19. Reverse fold.

20. Valley fold and unfold along edge of white triangle.

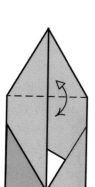
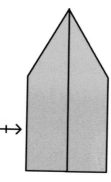
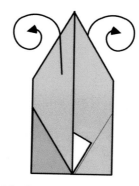
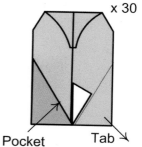

Finished Unit

x 30

21. Valley fold and unfold.

22. Repeat Steps 18–21.

23. Curl front tip toward you and back tip away from you with a narrow pencil-like object.

Pocket   Tab

Note that the other tab-pocket pair is at the back.

Locking

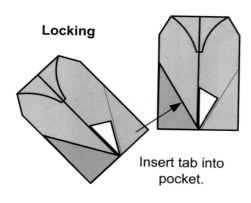

Insert tab into pocket.

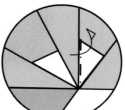

Firmly mountain fold the two units together toward the left as shown in the enlarged inset, thus locking the units. Tuck the folded layers underneath.

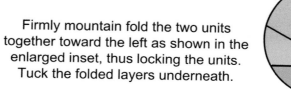

Two units locked.

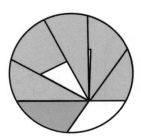

Assembly

Assemble 30 units in an icosidodecahedral symmetry as explained in the beginning of the chapter.

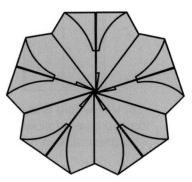

One finished face.

Note that you may also try the exposed lock method of the previous model, the Daylily.

Waterlily model made with gold and brown gift wrap paper.

Layered Poinsettia

Obtaining 1:3.5 Rectangles

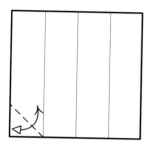

1. Cupboard fold and unfold to divide paper into fourths. Valley fold and unfold one corner.

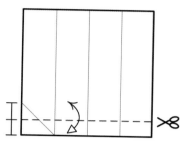

2. Valley fold and unfold as shown, then cut along new crease.

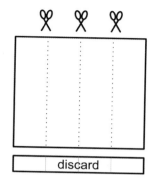

3. Cut along creases made in Step 1 to obtain 1:3.5 strips.

discard

1:3.5

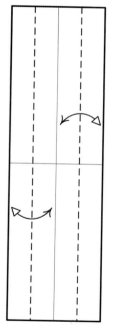

1. Crease both book-folds, then cupboard fold and unfold.

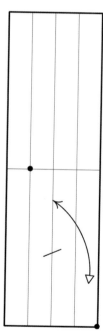

2. Match dots to pinch on the book-fold.

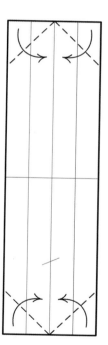

3. Valley fold corners.

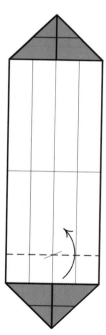

4. Valley fold along reference point made in Step 2.

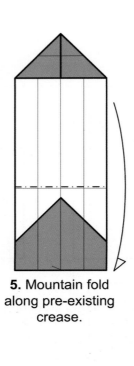

5. Mountain fold along pre-existing crease.

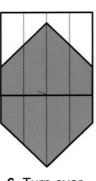

6. Turn over.

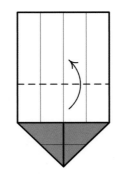

7. Valley fold to match length of layer underneath.

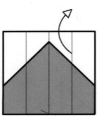

8. Unfold as shown.

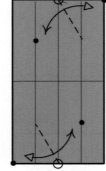

9. Valley fold to match dots and unfold.

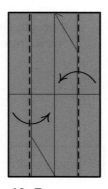

10. Re-crease cupboard folds.

11. Valley fold tracing creases at the back and unfold.

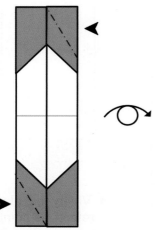

12. Reverse fold and turn over.

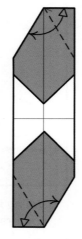

13. Valley fold and unfold.

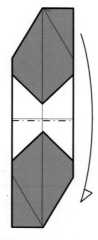

14. Mountain fold along pre-existing crease.

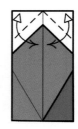

15. Valley fold corners and unfold.

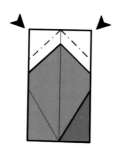

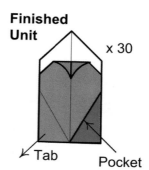

16. Reverse fold.

17. Curl colored petal toward you with a narrow pencil-like object. Repeat at the back.

Finished Unit

x 30

Tab

Pocket

Note that the other tab-pocket pair is at the back.

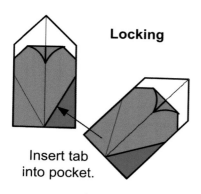

Locking

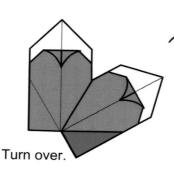

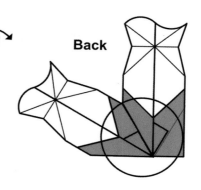

Back

Insert tab into pocket.

Turn over.

Firmly valley fold the two units together toward the left as shown in the enlarged inset, thus locking the units.

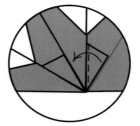

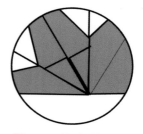

Two units locked.

Assembly

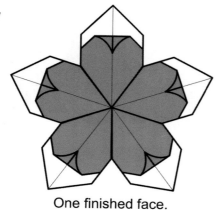

One finished face.

Assemble 30 units in an icosidodecahedral symmetry exactly as explained in the beginning of the chapter.

Two views of Layered Poinsettia made with simulated duo matte foils.

Daisy

The Daisy model is almost identical to the Layered Poinsettia model, except that certain folds are in reverse. Do Steps 1–12 of Layered Poinsettia on page 91, but make sure to start with *colored side of the paper facing you.* Continue with the rest of the steps below.

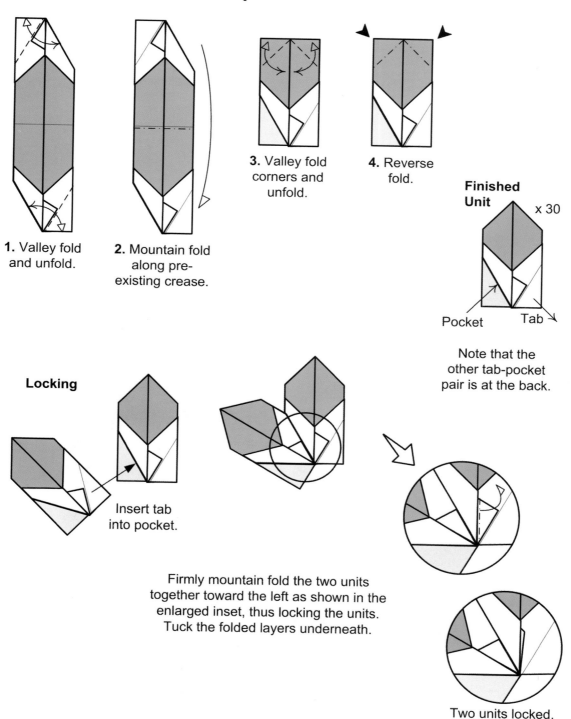

1. Valley fold and unfold.

2. Mountain fold along pre-existing crease.

3. Valley fold corners and unfold.

4. Reverse fold.

Finished Unit
x 30

Pocket   Tab

Note that the other tab-pocket pair is at the back.

Locking

Insert tab into pocket.

Firmly mountain fold the two units together toward the left as shown in the enlarged inset, thus locking the units. Tuck the folded layers underneath.

Two units locked.

Assembly

Assemble 30 units in an
icosidodecahedral symmetry
as explained in the
beginning of the chapter.

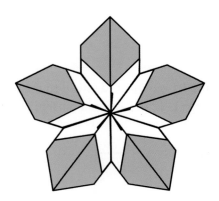

One finished face.

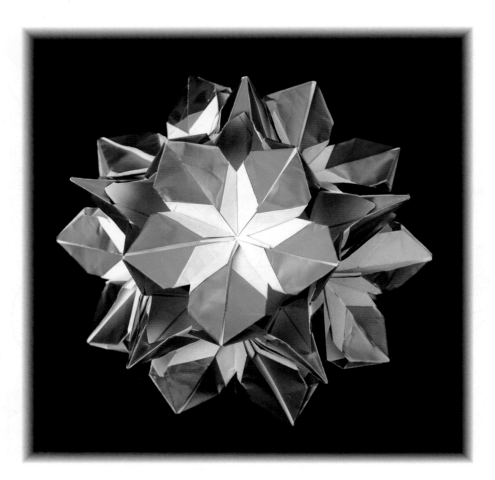

Daisy made with matte foil paper.

Layered Passion Flower

Start with 1:4 rectangles. Many steps are identical to the Layered Poinsettia model on page 91 and you must refer to it as directed.

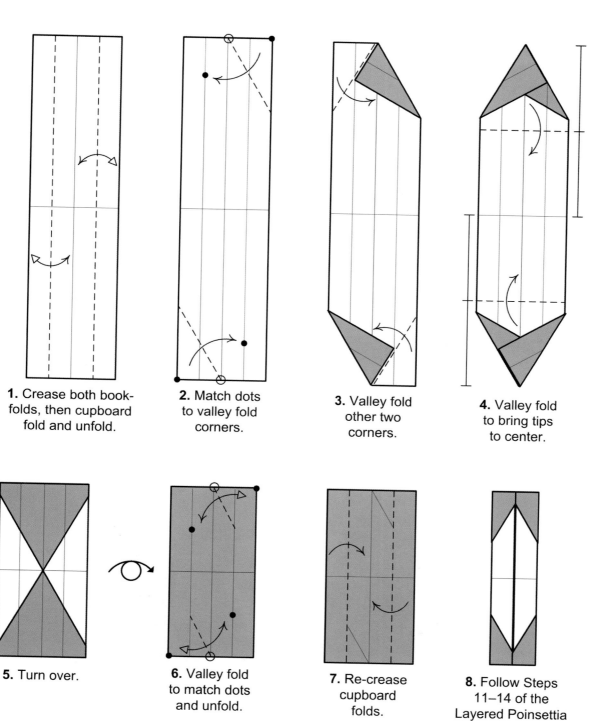

1. Crease both book-folds, then cupboard fold and unfold.

2. Match dots to valley fold corners.

3. Valley fold other two corners.

4. Valley fold to bring tips to center.

5. Turn over.

6. Valley fold to match dots and unfold.

7. Re-crease cupboard folds.

8. Follow Steps 11–14 of the Layered Poinsettia model.

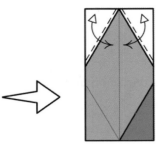

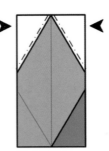

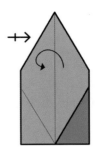

9. Valley fold along edges shown and unfold.

10. Reverse fold. The corners will overlap underneath.

11. Curl front tip toward you with a narrow pencil-like object. Repeat at the back.

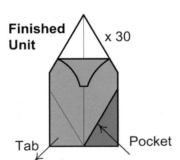

Finished Unit × 30

Tab

Pocket

Note that the other tab-pocket pair is at the back.

Assembly

Lock and assemble 30 units in an icosidodecahedral symmetry exactly as explained in the Layered Poinsettia model.

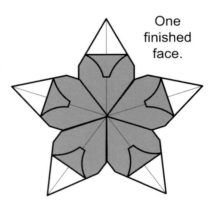

One finished face.

Layered Passion Flower Ball made with simulated duo paper using inserts.

Layered Petunia

Start with 1:4 rectangles. Many steps are identical to the Layered Poinsettia model on page 91 and you should refer to it as directed.

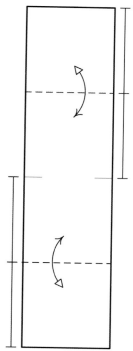

1. Pinch ends of horizontal book-folds, then valley fold and unfold as shown.

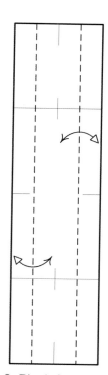

2. Pinch four marks along book-fold as shown, then cupboard fold and unfold.

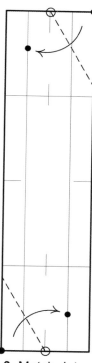

3. Match dots to valley fold corners.

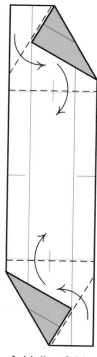

4. Valley fold other two corners, then re-crease folds from Step 1.

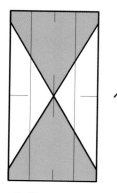

5. Turn over.

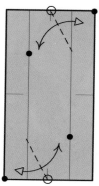

6. Valley fold to match dots and unfold.

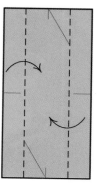

7. Re-crease cupboard folds.

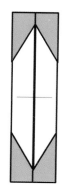

8. Follow Steps 11–14 of the Layered Poinsettia model.

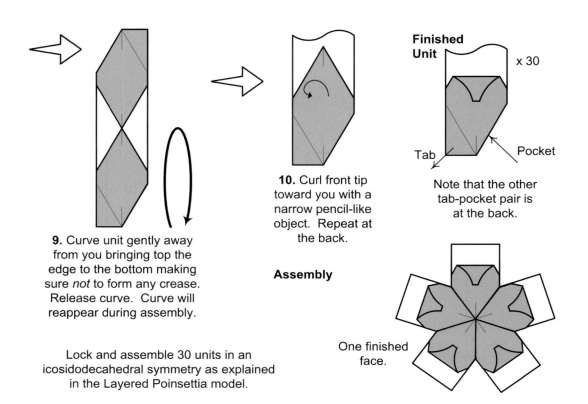

9. Curve unit gently away from you bringing top the edge to the bottom making sure *not* to form any crease. Release curve. Curve will reappear during assembly.

10. Curl front tip toward you with a narrow pencil-like object. Repeat at the back.

Finished Unit

x 30

Tab

Pocket

Note that the other tab-pocket pair is at the back.

Assembly

Lock and assemble 30 units in an icosidodecahedral symmetry as explained in the Layered Poinsettia model.

One finished face.

Layered Petunia made with simulated duo paper using inserts.

Dianthus

Start with 1:3 rectangles. (Please see *Origami Symbols and Bases* in Chapter 1 to crease a square into thirds and thus obtain 1:3 rectangles.) Many steps will be identical to the Layered Poinsettia model on page 91 and you must refer to it as directed.

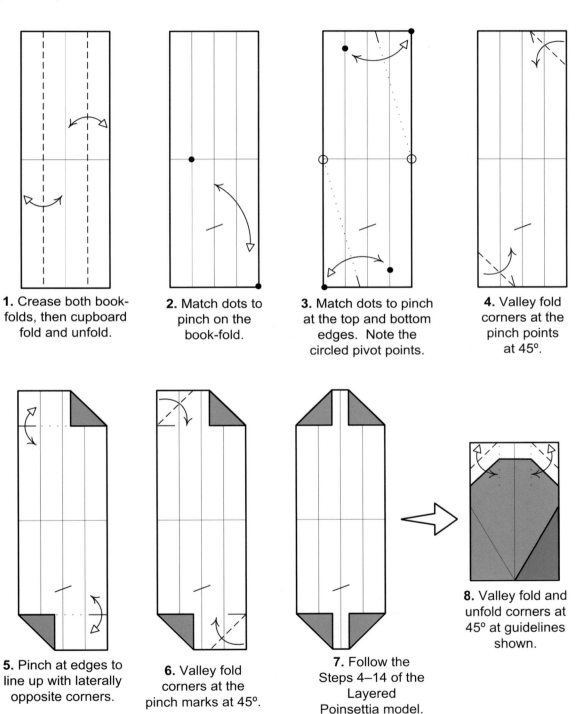

1. Crease both book-folds, then cupboard fold and unfold.

2. Match dots to pinch on the book-fold.

3. Match dots to pinch at the top and bottom edges. Note the circled pivot points.

4. Valley fold corners at the pinch points at 45°.

5. Pinch at edges to line up with laterally opposite corners.

6. Valley fold corners at the pinch marks at 45°.

7. Follow the Steps 4–14 of the Layered Poinsettia model.

8. Valley fold and unfold corners at 45° at guidelines shown.

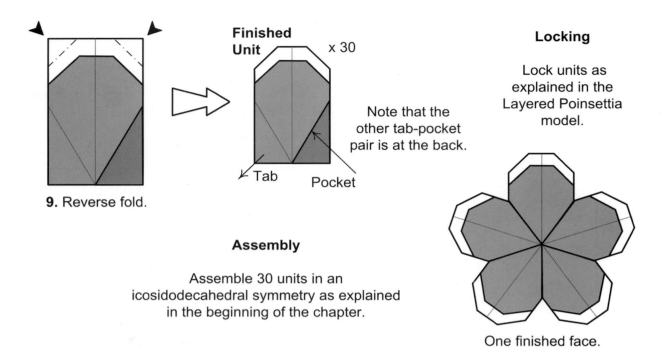

9. Reverse fold.

Finished Unit x 30

Note that the other tab-pocket pair is at the back.

Tab   Pocket

Locking

Lock units as explained in the Layered Poinsettia model.

Assembly

Assemble 30 units in an icosidodecahedral symmetry as explained in the beginning of the chapter.

One finished face.

The Dianthus model made with matte foil paper.

Hexagonal Ornaments

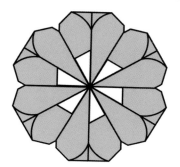

Daylily ornament.

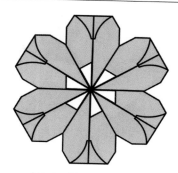

Waterlily ornament
with exposed locks.

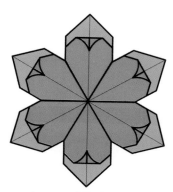

Layered Poinsettia
ornament in duo paper.

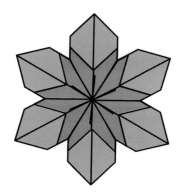

Daisy ornament in
duo paper.

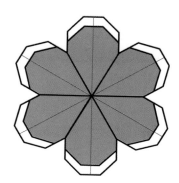

Dianthus ornament.

Layered Passion
Flower ornament.

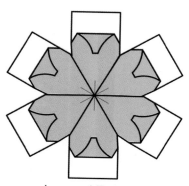

Layered Petunia
ornament.

The figures show six-unit hexagonal ornaments made of each of the
models in this chapter. Connect six units in a ring. Then turn over and
connect the six unconnected tabs and pockets to complete the ornament.

TUVWXYZ Stars (top) and QRSTUVWXYZ Stars (bottom) Planar Models.

7 ◆ Planar Models

The models in this chapter are comprised of intersecting planes. The first model in this genre of origami was the XYZ model (three intersecting planes) by Ed Sullivan [Ran69]. However, it was not until Tung Ken Lam created his WXYZ model (four intersecting planes) [Bos01] that a whole series of models with larger numbers of intersecting planes followed. Suddenly, people from all over the world began to create planar models in the earlier part of this decade.

The advent of planar models is quite interesting and world renowned origami artist David Petty has done extensive research on it. He tracked down and pursued many planar model creators for answers and missing links. He designed a few planar models himself and encouraged me to create more of these models after I had designed a handful. Petty's research, along with his exhaustive list of model drawings has, been presented in the next section, the *History of Planar Models*, with his kind permission. Although there are over 40 models in this series, diagrams for how to fold only the ten that are my designs are presented in this chapter. Of the rest of the models, most are still unpublished to this date.

As mentioned, David Petty is an origami artist known worldwide. A scientist by profession, he has created many origami models. He has contributed to origami exhibitions around the world and has taught origami in many countries. His work can be found in several books and numerous origami society magazines and publications, including those of the British Origami Society. Some of his work can also be found on his website is http://members.aol.com/ukpetd.

Photocopy paper works best for these models. The thickness is perfect and the models end up being sturdy. There is no requirement for a contrasting color for the back side of the paper because the back side does not get exposed. *Kami* works well too but is not a requirement. The star class models are made with square paper and the rectangular-arm class models are made with 2:1 rectangles. For the latter, rectangles of larger aspect ratio may also be used.

Recommendations

Paper Size: 4″ squares for star class, 3″x6″ rectangles for others

Paper Type: Photocopy paper or *kami*

Finished Model Size (approximate)

Star Class: 4″ squares yield models of height 5.6″.

Other: 3″x6″ rectangles yield models of height 6″.

History of Planar Models

David Petty

Planar Modular Pieces

The recent story begins with WXYZ which was created by Tung Ken Lam. This piece was first exhibited at the British Origami Society (BOS) convention in April 2001; later diagrams appeared in the BOS magazine (June 2001) and on David Petty's web site [Pet]. Tung Ken told everyone that the model was inspired by Ed Sullivan's XYZ model that had appeared in the 1970s along with a piece it had inspired, the Omega Star by Philip Shen. Along with these models, also included in the list of figures on the following pages, are other earlier models by Tung Ken Lam, Octahedral Cross and Blintz Icosidodecahedron. These were his antecedents. Good though they may be, neither had the impact that the WXYZ was to have.

The cudgels were first taken up by Francis Ow, who produced a whole series of XYZ models as well as other WXYZ variations and in 2001, the first VWXYZ. Also in 2001, a number of other British folders (Mick Guy, David Petty and Nick Robinson) came up with other XYZ models. A new variant of WXYZ, a four-unit version, was produced by Daniel Kwan in 2000.

The first UVWXYZ appears to have come from Japan (Satoshi Kamiya and Ushio Ikegami) or the US (Daniel Kwan) in 2001. This was followed by other UVWXYZ models from Mark Leonard, Jeanine Mosely, and Meenakshi Mukerji in 2002. The first TUVWXYZ came from Leonard in 2003. This was followed by two more from Mukerji.

An offshoot from the planar nature of the pieces came from David Petty. Using Francis Ow's unit for XYZ, he produced a curious continuous spiral form christened Tornado. This model was exhibited at the BOS convention in 2002. A group of British folders (Mark Leonard, John McKeever, and Martin Gibbs) spent some time investigating the Tornado and eventually drew it to the attention of Robert Lang. Ow's unit was subsequently refined in 2003 by Lang, using computer modeling techniques to give a better fitting unit. In its turn, the folding method for Lang's unit was refined by Ian Harrison in 2003. That same year, Meenakshi Mukerji suggested an arrangement of colored units for Tornado, to emphasize the spiral nature of the piece. Also in 2003, Leonard, while searching for a better version of the Tornado, came up with a double spiral version and also a UVWXYZ. QRSTUVWXYZ came from Mukerji in mid-2003. This was quickly followed by STUVWXYZ and RSTUVWXYZ from the same creator.

Other related models were the "3D solids" by Ian Harrison. So, three years after it first appeared, there are more than three dozen members of the WXYZ-inspired family and it is still attracting attention.

Names

The distinctive naming convention used for all the planar modular pieces is based on geometric principles. The number of letters of the alphabet used denotes the number of planes, thus XYZ has 3 planes, WXYZ has 4 planes, and so on.

A secondary convention (though not strictly adhered to) is that the number of letters of the alphabet used, less one, denotes the number of sides of each plane, i.e., WXYZ has (4 – 1) = 3 for triangular planes. Some later pieces have sometimes used less rigorously defined shapes e.g., multipoint stars or multiarmed crosses (the latter have been called rectangles in most cases). In each case, the number of star points or arms of the cross match the number of sides.

Where the pieces have a nonplanar character, more distinctive names have been used, e.g., Tornado.

The following is a complete list of all known planar models, their predecessors, and their variants, as compiled by David Petty. The models in the list appear roughly in chronological order.

1. XYZ
3 intersecting squares
6 units

Ed Sullivan
(USA)

7. XYZ
3 intersecting rectangles
3 units

Mick Guy
(UK)

2. Omega Star
6 units

Philip Shen
(Hong Kong)

8. XYZ-1
3 intersecting squares
6 units

Francis Ow
(Singapore)

3. Octahedral Cross
3 intersecting crosses
12 units

Tung Ken Lam
(UK)

9. XYZ-2
3 intersecting squares
6 units

Francis Ow
(Singapore)

4. Blintz Icosidodecahedron
6 intersecting planes
30 units

Tung Ken Lam
(UK)

10. XYZ-Radiometer
3 intersecting blades
6 units

David Petty
(UK)

5. WXYZ
4 intersecting equilateral triangles
12 units

Tung Ken Lam
(UK)

11. XYZ-MBE
3 intersecting patterned squares
6 units

Nick Robinson
(UK)

6. XYZ
3 intersecting rectangles
6 units

Francis Ow
(Singapore)

12. XYZ-Diamonds
3 intersecting diamonds
6 units

Francis Ow
(Singapore)

13. XYZ-Diamonds
3 intersecting
diamonds
6 units

David Petty
(UK)

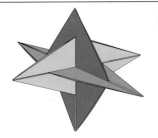

18. Modified Tweaked Tornado
Continuous single spiral
16 units

Ian Harrison (UK)
Robert Lang unit
(USA)

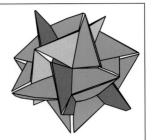

14. VWXYZ
5 intersecting
crosses
20 units

Francis Ow
(Singapore)

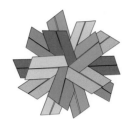

19. Tornado 2
Two spirals
20 units

Mark Leonard (UK)
Robert Lang unit
(USA)

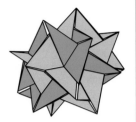

15. WXYZ-4 Piece
4 intersecting
equilateral triangles
4 units

Daniel Kwan
(USA)

20. WXYZ-Rectangles
4 intersecting 3 arm
crosses
12 units

Francis Ow
(Singapore)

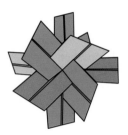

16. Tornado
Continuous single
spiral
16 units

David Petty (UK)
Francis Ow unit
(Singapore)

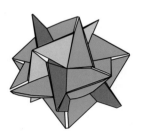

21. VWXYZ-Squares #1
5 intersecting square
planes
20 units

Francis Ow
(Singapore)

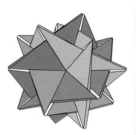

17. Tweaked Tornado
Continuous single
spiral
16 units

Robert Lang
(USA)

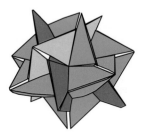

22. VWXYZ-Squares #2
5 intersecting square
Planes
20 units

Francis Ow
(Singapore)

23. VWXYZ-Squares
5 intersecting square
planes (distorted)
20 units

Mark Leonard (UK)
Francis Ow unit
(Singapore)

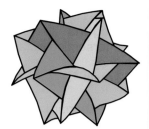

28. UVWXYZ-Rectangles
6 intersecting
5 arm crosses
30 units

Meenakshi Mukerji
(USA)

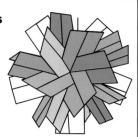

**24. UVWXYZ-
Pentagons**
6 intersecting
pentagonal
planes
30 units

Mark Leonard
(UK)

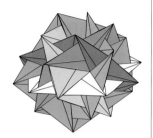

**29. TUVWXYZ-
Rectangles**
7 intersecting
6 arm crosses
42 units

Meenakshi Mukerji
(USA)

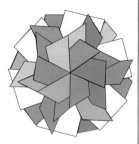

25. UVWXYZ-Stars
6 intersecting 5 point
star planes
30 units

Kamiya Satoshi
Ushio Ikegami
(Japan)

**30. TUVWXYZ-
Hexagons**
7 intersecting
hexagonal planes
42 units

Meenakshi Mukerji
(USA)

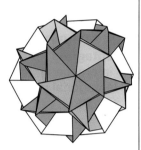

26. UVWXYZ-Pentagons
6 intersecting pentagonal
planes
30 units

Daniel Kwan
(USA)

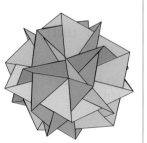

31. TUVWXYZ-Stars
7 intersecting
6 point star planes
42 units

Meenakshi Mukerji
(USA)

27. UVWXYZ-Stars
6 intersecting 5 point star
planes
30 units

Jeannine Mosely
(USA)

**32. TUVWXYZ-
Hexagons**
7 intersecting
Hexagonal planes
42 units

Mark Leonard
(UK)

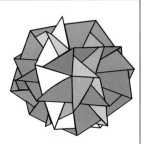

33. STUVWXYZ-Stars
8 intersecting 7 point stars
56 units

Meenakshi Mukerji
(USA)

34. STUVWXYZ-Rectangles
8 intersecting 7 point stars
56 units

Meenakshi Mukerji
(USA)

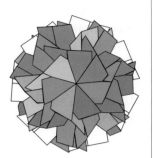

35. STUVWXYZ-Heptagons
8 intersecting Heptagonal planes
56 units

Meenakshi Mukerji
(USA)

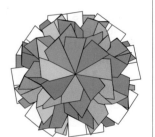

36. RSTUVWXYZ-Stars
9 intersecting 8 point stars
72 units

Meenakshi Mukerji
(USA)

37. RSTUVWXYZ-Rectangles
9 intersecting 8 arm crosses
72 units

Meenakshi Mukerji
(USA)

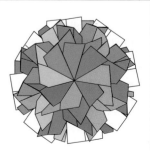

38. RSTUVWXYZ-Octagons
9 intersecting octagonal planes
72 units

Mark Leonard
(UK)

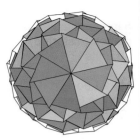

39. QRSTUVWXYZ-Stars
10 intersecting 9 point stars
90 units

Meenakshi Mukerji
(USA)

**40. 72° UNIT
= UVWXYZ**
30 units

Ian Harrison
(UK)

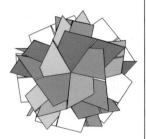

41. 90° UNIT
20 units

Ian Harrison
(UK)

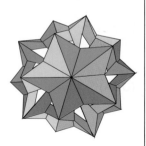

42. 120° UNIT
12 units

Ian Harrison
(UK)

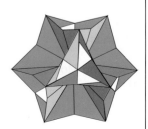

UVWXYZ Rectangles

Start with 1:2 paper and pinch ends of book-fold.

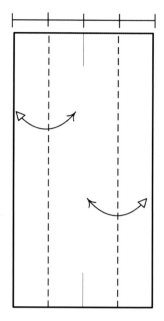

1. Cupboard fold and unfold.

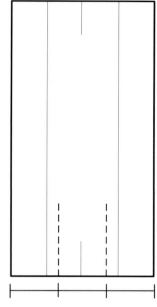

2. Make 1/3 creases as shown.

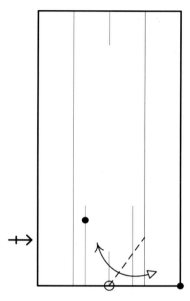

3. Match dots to form new crease as shown, then repeat on the left.

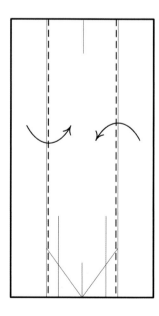

4. Re-crease cupboard folds from Step 1.

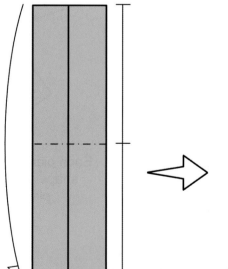

5. Mountain fold into half.

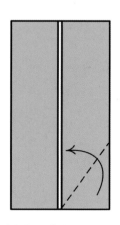

6. Valley fold to trace pre-existing crease at the back.

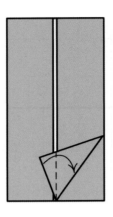

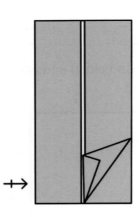

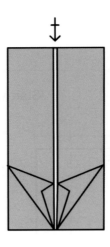

7. Valley fold along center line.

8. Repeat Steps 6 and 7 on the left.

9. Repeat Steps 6–8 at the back.

Finished Unit

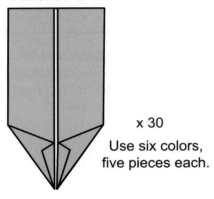

x 30

Use six colors, five pieces each.

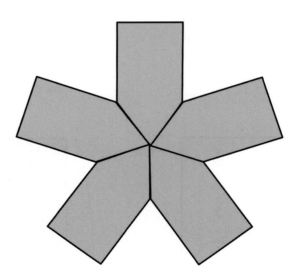

Each plane in the finished model will have a shape as above. There will be six such planes intersecting each other.

Assembly

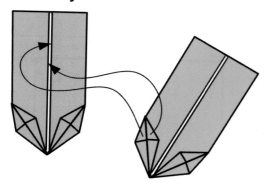

Open bottom flaps out slightly. Two flaps lying on one edge of a unit together form a tab and the opening running at the center is a pocket. Insert tab inside pocket as shown. Then gently slide second unit to align with the first unit at the bottom.

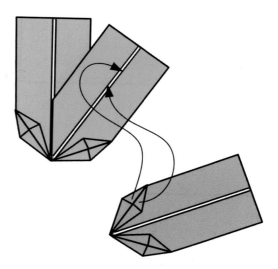

Use five different colors to make a ring of five. The tab of the first unit goes inside the pocket of the fifth unit to complete the ring. Extend each color to make six intersecting planes, the sixth color being added halfway.
Hint: During assembly think of an icosidodecahedron. The completed model has 12 rings of five and 20 rings of three.

TUVWXYZ Stars

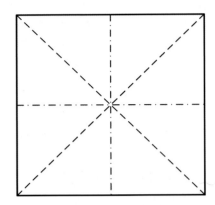

1. Fold like waterbomb base and unfold.

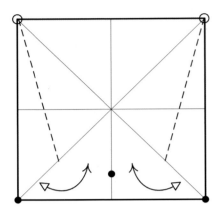

2. Crease to match dots, creasing only the parts shown.

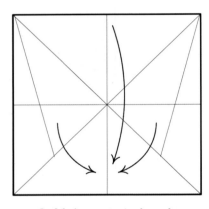

3. Make a waterbomb base.

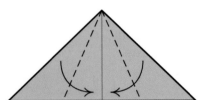

4. Fold to bring edges to center.

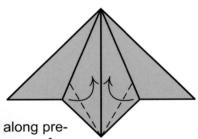

5. Fold along pre-existing crease from Step 2.

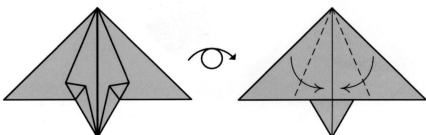

6. Turn around.

7. Repeat Step 4.

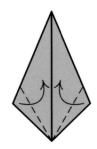

8. Valley fold to match edges at the back.

Finished Unit

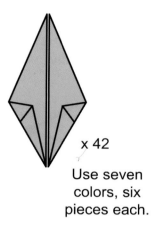

x 42

Use seven
colors, six
pieces each.

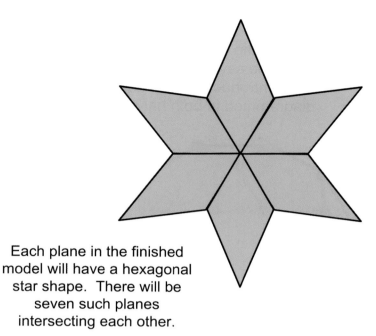

Each plane in the finished
model will have a hexagonal
star shape. There will be
seven such planes
intersecting each other.

Assembly

Open bottom flaps out slightly. Two flaps lying on
one edge of a unit together form a tab and the
opening running at the center is a pocket. Insert tab
inside pocket as shown. Then gently slide second
unit to align with the first unit at the bottom.

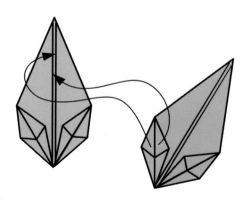

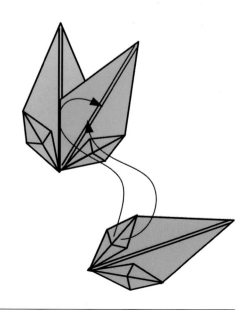

Use six different colors to make a ring of six. The tab of the first unit goes inside the pocket of the sixth unit to complete the ring. Refer to the color chart below to complete the seven intersecting rings of six, with the seventh color (shown as black) introduced about halfway during assembly. The seventh color is diagrammed in both halves to illustrate the continuity of the assembly.

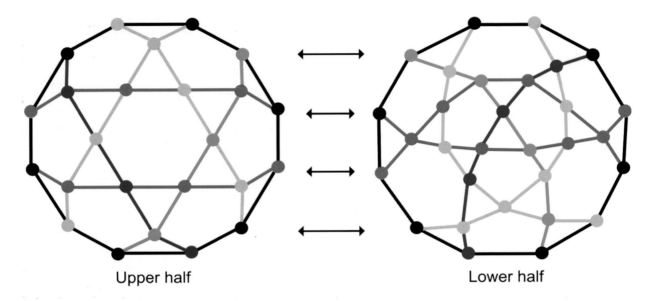

Upper half   Lower half

●———— is the diagrammatic representation of one unit.
Due to the abstract nature of the chart, a unit may appear bent.

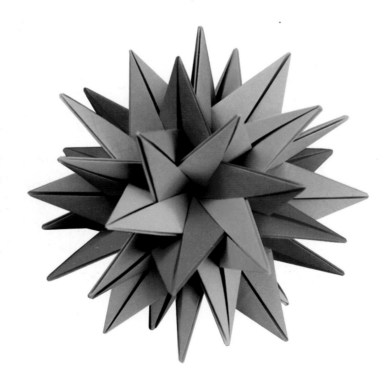

TUVWXYZ Rectangles

Start with 1:2 paper and pinch ends of book-fold.

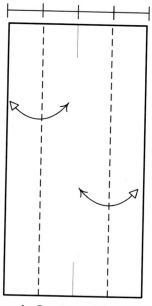

1. Cupboard fold and unfold.

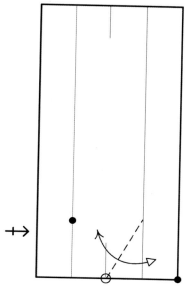

2. Match dots to form new crease as shown, then repeat on left.

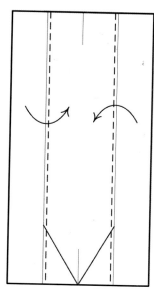

3. Re-crease cupboard folds.

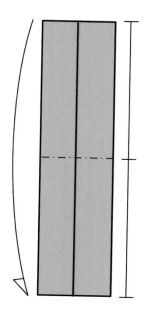

4. Mountain fold into half.

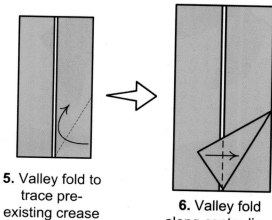

5. Valley fold to trace pre-existing crease at the back.

6. Valley fold along center line.

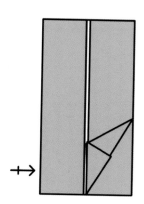

7. Repeat Steps 5 and 6 on the left.

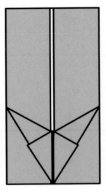 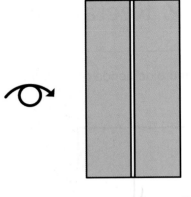 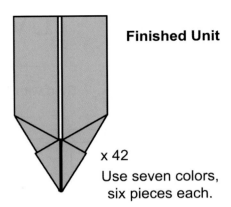

Finished Unit

8. Turn around.

9. Repeat Steps 5–8 by matching edges behind.

x 42
Use seven colors, six pieces each.

Assembly

Assemble as TUVWXYZ Stars. Each plane will appear as shown on the right.

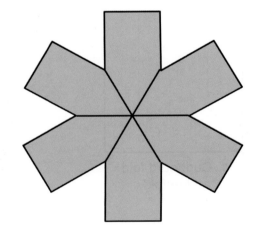

TUVWXYZ Hexagons

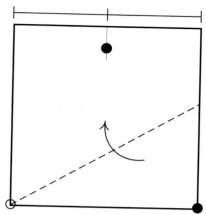

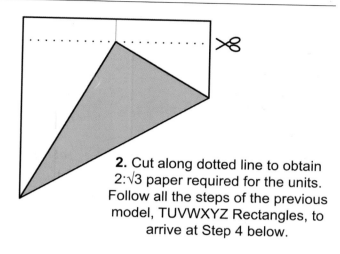

1. Pinch center of top edge and then match dots to crease as shown.

2. Cut along dotted line to obtain 2:√3 paper required for the units. Follow all the steps of the previous model, TUVWXYZ Rectangles, to arrive at Step 4 below.

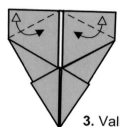

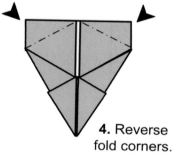

Finished Unit

3. Valley fold and unfold.

4. Reverse fold corners.

x 42

Use seven colors, six pieces each.

Assembly

Assemble like other TUVWXYZ models. Each plane will be of hexagonal shape.

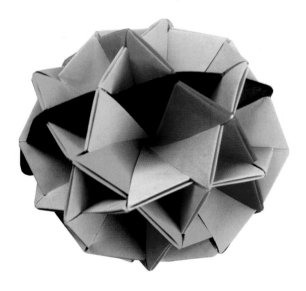

STUVWXYZ Stars

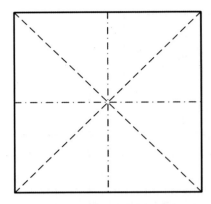

1. Fold like waterbomb base and unfold.

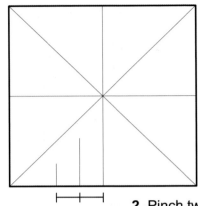

2. Pinch two halfway points as shown.

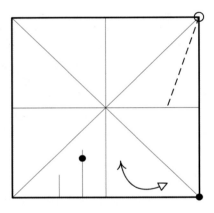

3. Fold to match dots and unfold.

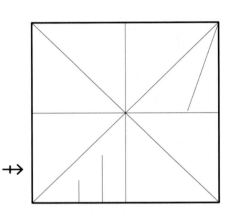

4. Repeat Steps 2 and 3 on the left.

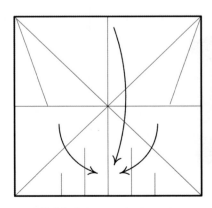

5. Make a waterbomb base.

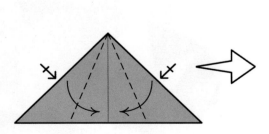

6. Crease edges to center. Repeat at the back.

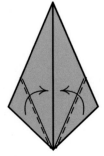

7. Fold along pre-existing creases.

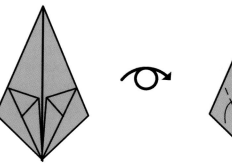
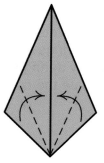

8. Turn around.

9. Fold to match edges at the back.

Finished Unit

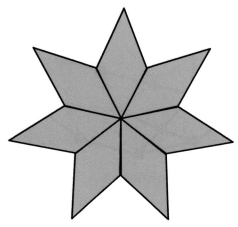

x 56

Use eight colors, seven pieces each.

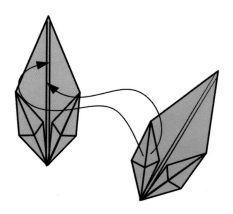

Each plane in the finished model will have a heptagonal star shape. There will be eight such planes intersecting each other.

Assembly

Open bottom flaps out slightly. Two flaps lying on one edge of a unit together form a tab and the opening running at the center is a pocket. Insert tab inside pocket as shown. Then gently slide second unit to align with the first unit at the bottom.

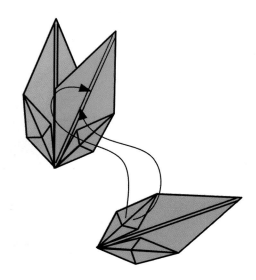

Use seven different colors to make a ring of seven. The tab of the first unit goes inside the pocket of the seventh unit to complete the ring. Refer to the color chart below to complete the eight intersecting rings of seven, with the eighth color (shown as black) introduced about halfway during assembly. The eighth color is diagrammed in both halves to illustrate the continuity of the assembly.

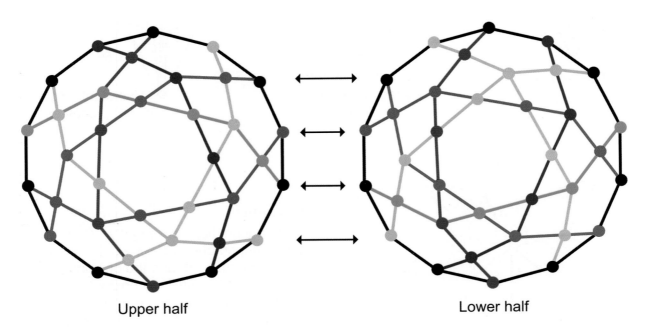

Upper half   Lower half

━━━━●━━━━ is the diagrammatic representation of one unit.
Due to the abstract nature of the chart, a unit may appear bent.

STUVWXYZ Rectangles

Start with 1:2 paper and pinch ends of book-fold.

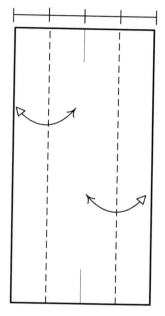

1. Cupboard fold and unfold.

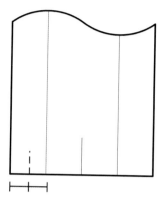

2. Pinch halfway point of bottom edge as shown.

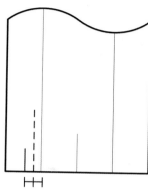

3. Crease another halfway point as shown.

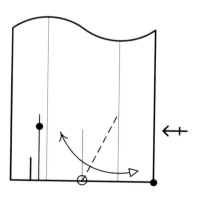

4. Match dots to form new crease as shown, then repeat Step 2–4 on the right.

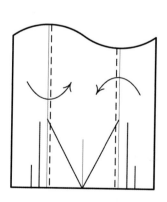

5. Re-crease cupboard folds.

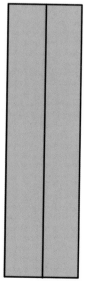

6. Repeat Steps 4–6 of TUVWXYZ Rectangles.

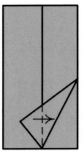 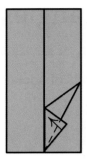 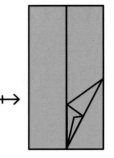 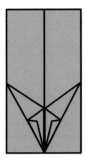

7. Crease
along center.

8. Fold in
extra.

↦ **9**. Repeat
Steps 7 and
8 on the left.

10. Turn around and
finish the back like the
front by matching edges
in the front.

Finished Unit

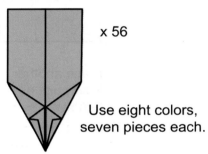

x 56

Use eight colors,
seven pieces each.

Assembly

Assemble as
STUVWXYZ Stars.
Each plane will appear
as shown on the right.

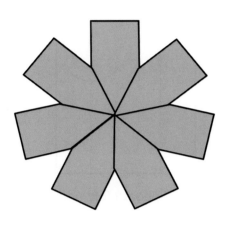

STUVWXYZ Heptagons

1. Start with 8.5"x11" standard US letter paper and cut into quarters as shown. Follow all the steps of the previous model, the STUVWXYZ Rectangles, to arrive at the next step.

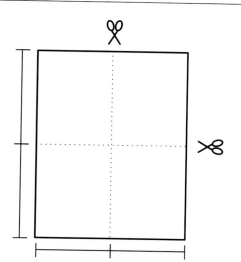

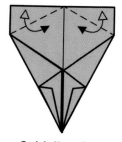

2. Valley fold and unfold.

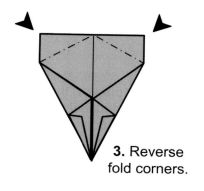

3. Reverse fold corners.

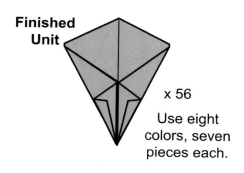

Finished Unit

x 56

Use eight colors, seven pieces each.

Assembly
Assemble like STUVWXYZ Stars. Each plane will be of heptagonal shape.

RSTUVWXYZ Stars

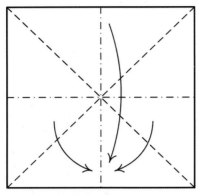

1. Make a waterbomb base.

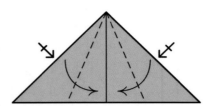

2. Fold edges to center. Repeat at the back.

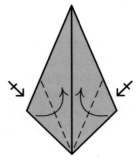

3. Fold bottom edges to center and repeat at the back.

Finished Unit

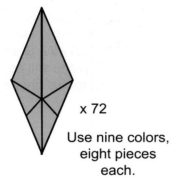

x 72

Use nine colors, eight pieces each.

Each plane in the finished model will have an octagonal star shape. There will be nine such planes intersecting each other.

Assembly

Open bottom flaps out slightly. Two flaps lying on one edge of a unit together form a tab and the opening running at the center is a pocket. Insert tab inside pocket as shown. Then gently slide second unit to align with the first unit at the bottom.

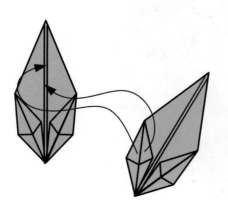

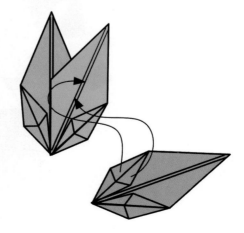

Use eight different colors to make a ring of eight. The tab of the first unit goes inside the pocket of the eighth unit to complete the ring. Refer to the color chart below to complete the nine intersecting rings of eight, with the ninth color (shown as black) introduced about halfway during assembly. The ninth color is diagrammed in both halves to illustrate the continuity of the assembly.

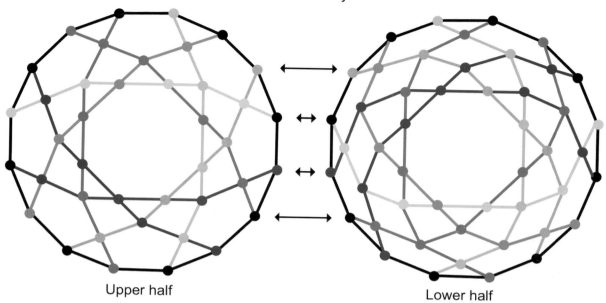

Upper half   Lower half

──●── is the diagrammatic representation of one unit.
Due to the abstract nature of the chart, a unit may appear bent.

RSTUVWXYZ Rectangles

Start with 1:2 paper and pinch ends of book-fold.

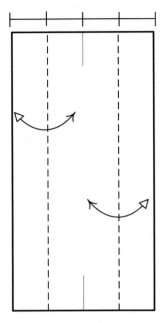

1. Cupboard fold and unfold.

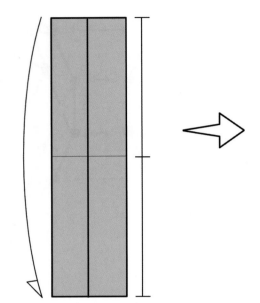

2. Mountain fold into half.

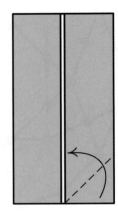

3. Valley fold to bring bottom right to center.

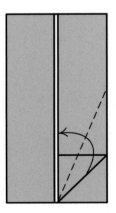

4. Valley fold to bring new edge to center.

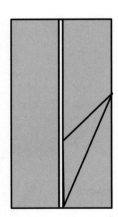

5. Unfold to Step 3.

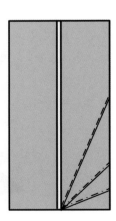

6. Valley and mountain fold pre-existing creases as shown.

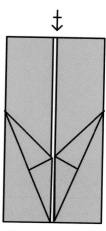

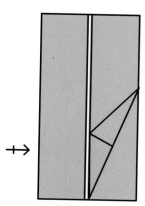

7. Repeat Steps 3–6 on the left.

8. Repeat Steps 3–7 at the back.

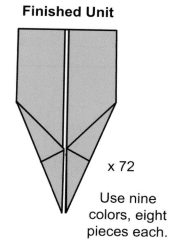

Finished Unit

x 72

Use nine colors, eight pieces each.

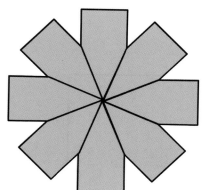

Assembly

Each plane in the finished model will appear as shown on the left. There will be nine such planes intersecting each other. Assemble like RSTUVWXYZ Stars.

QRSTUVWXYZ Stars

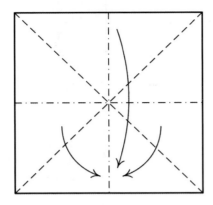

1. Make a waterbomb base.

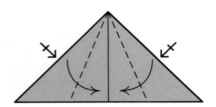

2. Crease edges to center. Repeat at the back.

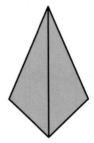

3. Unfold all the way.

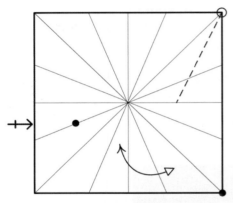

4. Fold and unfold to match dots. Repeat on left.

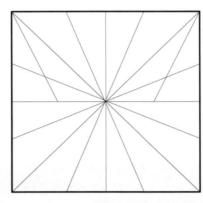

5. Repeat Steps 1 and 2.

6. Valley fold along pre-existing crease.

 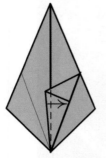

7. Valley fold upper flap along center line.

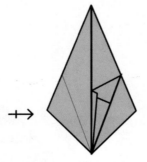

8. Repeat steps 6 and 7 on the left and then match the back to the front.

Finished Unit

x90

Use ten colors, nine pieces each.

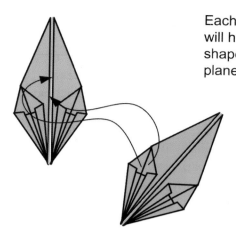

Each plane in the finished model will have a nanogonal star shape. There will be ten such planes intersecting each other.

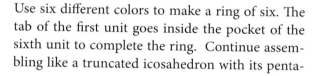

Assembly
Open bottom flaps out slightly. Two flaps lying on one edge of a unit together form a tab and the opening running at the center is a pocket. Insert tab inside pocket as shown. Then gently slide one unit to align with the other at the bottom.

Use six different colors to make a ring of six. The tab of the first unit goes inside the pocket of the sixth unit to complete the ring. Continue assembling like a truncated icosahedron with its penta- gons and hexagons rotated to meet at their verti- ces, thus introducing rings of three in the process. The completed model has 12 rings of five, 20 rings of six and 60 rings of three.

Mathematics of Planar Models

As pointed out by David Petty, the first origami planar model was the XYZ by Ed Sullivan [Ran69] created in the late 1960s. This planar construction was a paper model of the familiar three intersecting planes called the Cartesian planes. The model of the next degree in this series came much later in 2001 and was the WXYZ model by Tung Ken Lam [Bos01] which consisted of four intersecting planes. The naming scheme was such that each letter of the alphabet used represented one plane, just like its Cartesian planes predecessor. Although the four plane model came much later, it triggered a flurry of models of higher degrees consisting of five, six, seven, eight, nine and even ten intersecting planes. The planes themselves were being made of various shapes such as stars with pointed tips, stars with rectangular arms, and various polygons. Two examples of such planes are illustrated below.

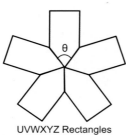

UVWXYZ Rectangles

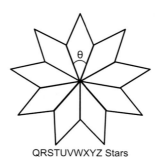

QRSTUVWXYZ Stars

The mathematics of the planar models is rather interesting. The two basic properties that have been observed are: (i) every plane intersects every other plane and (ii) all planes pass through one single point, the center of the model. Let us take a planar of degree n, which means that there are n intersecting planes. Since every plane intersects every other plane, every plane intersects $n - 1$ other planes.

Thinking in terms of origami, the lines where the planes intersect are basically the lines where the units meet and are attached together by means of tabs and pockets. Therefore, for a plane to intersect $n - 1$ other planes, there must be $n - 1$ joints on the plane, which in turn means that the planes themselves have to be made up of $n - 1$ units each since the units meet in a circle. Therefore, it is no mere coincidence that an origami planar model of degree n has $n - 1$ units per plane and is therefore made up of a total of $n(n - 1)$ units. For example, let us take the UVWXYZ model, which has six intersecting planes. The total number of units required to make a UVWXYZ is $6 \times (6 - 1) = 30$ units. Similarly, the total number of units required to make a QRSTUVWXYZ model which has ten intersecting planes is $10 \times (10 - 1) = 90$ units.

Now, looking at each plane, all $n - 1$ units meet at the center of the plane, which means that the angle of each unit at the center of the plane $\theta = 360°/(n - 1)$ and this is true, obviously, regardless of what shape we choose for the plane. Using the same examples as the above paragraph, the units of a UVWXYZ model would have $\theta = 360°/(6 - 1) = 72°$ and the units of a QRSTUVWXYZ model would have $\theta = 360°/(10 - 1) = 40°$. Knowing the value of θ is crucial to designing any unit of the planar class. In the designs we can be a fraction of a degree off from the mathematical value of θ but we must make sure that we err on the lesser side of θ, rather than on the greater side, because the latter will result in warped planes. Erring on the lesser side actually is advantageous because it allows for some correction due to the thickness of the paper. Also, the units themselves can open up a bit from their flat original forms to compensate for the negative error.

Planar Properties Summary

◈ Every plane passes through the center of the model.

◈ Every plane intersects every other plane.

◈ If a planar is of degree n, i.e., there are n planes, then the polygonal nature of each plane is of degree $n - 1$.

◈ Total number of units is $U = n(n - 1)$.

◈ Angle at the base of the units is $\theta = 360/(n - 1)$.

◈ Sum of number of sides of all facial polygons in the assembly is $\Sigma S = 4U$.

Model Name	Number of Planes (n)	Nature of Each Plane $(n-1)°$ Polygon	Total Number of Units $U = n(n-1)$	Base Angle $\theta = 360/(n-1)°$	Underlying Polyhedron in the Assembly	Facial Polygons Or Rings in the Assembly	Sum of Sides $\Sigma S = 4U$
XYZ (Cartesian Planes)	3	Digonal (2)	3 x 2 = 6	360/2 = 180°	Octahedron	8 (triangle)	24
WXYZ	4	Triangular (3)	4 x 3 = 12	360/3 = 120°	Cuboctahedron	8 (triangle), 6 (square)	24 + 24 = 48
VWXYZ	5	Quadrilateral (4)	5 x 4 = 20	360/4 = 90°	Pentagonal Gyrobicupola J31	10 (triangle), 10 (square), 2 (pentagon)	30 + 40 + 10 = 80
UVWXYZ	6	Pentagonal (5)	6 x 5 = 30	360/5 = 72°	Icosi-dodecahedron	20 (triangle), 12 (pentagon)	60 + 60 = 120
TUVWXYZ	7	Hexagonal (6)	7 x 6 = 42	360/6 = 60°	No name	28 (triangle), 12 (pentagon), 4 (hexagon)	84 + 60 + 24 = 168
STUVWXYZ	8	Heptagonal (7)	8 x 7 = 56	360/7 = 51.4°	No name	28 (triangle), 14 (square), 14 (hexagon), 2 (heptagon)	84 + 56 + 70 + 14 = 224
RSTUVWXYZ	9	Octagonal (8)	9 x 8 = 72	360/8 = 45°	No name	32 (triangle), 24 (square), 16 (hexagon), 2 (octagon)	96 + 96 + 80 + 16 = 288
QRSTUVWXYZ	10	Nanogonal (9)	10 x 9 = 90	360/9 = 40°	Truncated Icosahedron-like	60 (triangle), 12 (pentagon), 20 (hexagon)	180 + 60 + 120 = 360

(Note: For best results make each plane in one single color to emphasize the planes and also for ease of assembly. Disclaimer: In some of the planar models, especially those of higher degrees, the planes may not be perfect planes, i.e., they may be slightly distorted.)

Once the number of units in a model is determined, a planar assembly scheme needs to be planned. For some models this has been a simple task while for others it has been quite challenging. For those planar models that comprise of a number of units that have some relationship with the number of edges of a known polyhedron, the task is relatively simple, e.g., the XYZ, the WXYZ, and the UVWXYZ models are based on an octa-hedron, a cuboctahedron, and an icosidodecahedron respectively. The VWXYZ solution was first arrived at by Francis Ow. In this model the underlying polyhedron is the Johnson Solid named the Pentagonal Gyrobicupola (J31) [Joh66] which is not very common.

The first assembly based on an unusual polyhedron was the TUVWXYZ model and Mark Leonard [Bos03] was the one to come up with a solution. This assembly could possibly be based on a "near miss polyhedron" [Kap01] and has a tetrahedral symmetry. The assembly solutions for the rest of the planar models of higher degree were first arrived at by me. The STUVWXYZ and RSTUVWXYZ solutions were determined by trial and error. For the latter, I was inclined to think that it may have a relationship with the Archime-

dean solid rhombitruncated cuboctahedron (72 edges) but that solution did not seem to work. The QRSTUVWXYZ solution was based on a truncated icosahedron with the pentagons and hexagons resized and rotated if you will, such that the old vertices transformed into new triangles. I later came across this interesting solid as a "Near Miss Polyhedron" [Kap01].

Next, we will examine some of the planar models presented in this chapter and prove for each of these models that the origami methods used to achieve the desired θs discussed earlier are indeed what we were looking for in our designs. Understanding this might help the readers in the design of their own new planar models perhaps with different plane shapes, or in the design of new planar models of higher degrees. The proofs utilize simple principles from geometry and trigonometry.

UVWXYZ Rectangles

Required angle of a unit at the center of each plane
$\theta = 360/(6 - 1) = 360/5 = 72°$.

So, required φ as marked in the diagram = $\theta/2$ = 36°. Let the starting square be of side length a, so the size of the starting rectangle is $a \times a/2$.

Fold a unit as shown on page 111 up to Step 3 and extend the new crease all the way to the edge at C. Before unfolding, i.e., when B lies on Q, trace the lines OQ and QC along the edges of the little folded triangle. Hence, $\triangle OCQ \cong \triangle OCB$ as one is traced from the other.

In $\triangle POQ$, $OQ = OB = a/4$
and $OP = (a/2)/3 = a/6$
$\cos \beta = (a/6)/(a/2) = 1/3$.
Hence, $\beta = \cos^{-1}(1/3) = 70.5°$.

Since $\triangle OCQ \cong \triangle OCB$,
$\angle QOC = \angle BOC = (180 - \beta)/2 = 54.75°$.

Also, $\angle QOC = \varphi + (90 - \beta)$,
or $54.75 = \varphi + 90 - 70.5$.
Therefore, $\varphi = 35.25° \approx 36°$, a close enough approximation for this model.

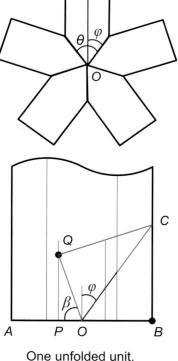

One of the six planes.

One unfolded unit.

TUVWXYZ Stars

Required angle of a unit at the center of each plane $\theta = 360/(7 - 1) = 360/6 = 60°$.

So, required φ as marked in the diagram $= \theta/2 = 30°$. Let the size of the starting square be of length a.

Fold a unit as shown on page 114 up to Step 2 and extend the new crease all the way to the edge at C. Before unfolding, i.e., when B lies on Q, trace the lines OQ and QC along the edges of the folded triangle. Hence, $\triangle OCQ \cong \triangle OCB$ as one is traced from the other.

In $\triangle OPQ$, $\cos(\angle POQ) = OP/OQ$
$= (a/2)/a = 1/2$.
Hence $\angle POQ = \cos^{-1}(1/2) = 60°$ and
$\angle QOB = 90 - \angle POQ = 90 - 60 = 30°$.

Since $\triangle OCQ \cong \triangle OCB$,
$\angle QOC = \angle BOC = \angle QOB/2 = 30/2 = 15°$.

Therefore, $\varphi = 45 - \angle BOC$
$= 45 - 15 = 30°$.

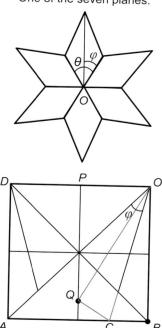

One of the seven planes.

One unfolded unit.

TUVWXYZ Rectangles

Required angle is the same as above, i.e., $\theta = 60°$. Let the size of the starting rectangle be $a \times a/2$.

Fold a unit as shown on page 117 up to Step 2 and extend the new crease all the way to the edge at C. Before unfolding, i.e., when B lies on Q, trace the lines OQ and QC along the edges of the folded triangle. Hence, $\triangle OCQ \cong \triangle OCB$ as one is traced from the other.

In $\triangle POQ$,
$OQ = OB = a/4$ and $OP = a/8$
$\cos \beta = (a/8)/(a/4) = 1/2$.
Hence, $\beta = \cos^{-1}(1/2) = 60°$.

Since $\triangle OCQ \cong \triangle OCB$,
$\angle QOC = \angle BOC = (180 - \beta)/2 = 60°$
$= (180 - 60)/2 = 60°$.
Therefore, $\theta = \angle QOC = 60°$.

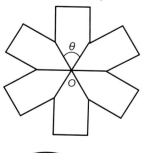

One of the seven planes.

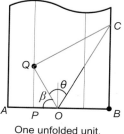

One unfolded unit.

STUVWXYZ Stars

Required angle of a unit at the center of each plane
$\theta = 360/(8 - 1) = 51.43°$.

So, required φ as shown in diagram = $\theta/2 = 25.71°$.
Let side of starting square be a.

Fold a unit as shown on page 120 up to Step 3 and extend the new crease all the way to the edge at C. Before unfolding, i.e., when B lies on Q, trace the lines OQ and QC along the edges of the folded triangle. Hence, $\Delta OCQ \cong \Delta OCB$ as one is traced from the other.

In ΔOPQ, $OQ = a$ and $OP = a/2 + a/8 = 5a/8$
$\cos \beta = OP/OQ = (5a/8)/a = 5/8$.
Hence, $\beta = \cos^{-1}(5/8) = 51.32°$.

$LAOQ = \beta - 45 = 51.32 - 45 = 6.32°$.
$LQOB = LAOB - LAOQ = 45 - 6.32 = 38.68°$.

Since $\Delta OCQ \cong \Delta OCB$, $LQOC = LBOC$
$= LQOB/2 = 38.68/2 = 19.34°$.
Therefore, $\varphi = LQOC + LAOQ$
$= 19.34 + 6.32 = 25.66° \approx 25.71°$, a close enough approximation for this model.

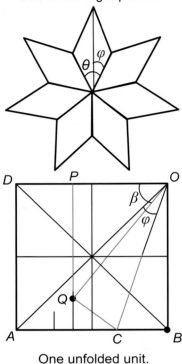

One of the eight planes.

One unfolded unit.

STUVWXYZ Rectangles

Required angle is the same as above, i.e., $\varphi = \theta/2 = 25.71°$. Let the size of the starting rectangle be a x $a/2$.

Fold a unit as shown on page 123 up to Step 4 and extend the new crease all the way to the edge at C. Before unfolding, i.e., when B lies on Q, trace the lines OQ and QC along the edges of the folded triangle. Hence, $\Delta OCQ \cong \Delta OCB$ as one is traced from the other.

In ΔOPQ,
$OQ = a/4$ and $OP = a/8 + a/32 = 5a/32$.
$\cos \beta = OP/OQ = (5a/32)/(a/4) = 5/8$.
Hence, $\beta = \cos^{-1}(5/8) = 51.32°$.

Since $\Delta OCQ \cong \Delta OCB$, $LQOC = LBOC$
$= LQOB/2 = (180 - \beta)/2 = 64.34°$.

Also, $LQOC = \varphi + (90 - \beta)$
or $64.34 = \varphi + 90 - 51.32$.
Therefore, $\varphi = 25.66° \approx 25.71°$, a close enough approximation for this model.

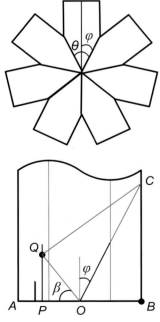

One of the eight planes.

One unfolded unit.

RSTUVWXYZ Stars and RSTUVWXYZ Rectangles

Shown on the right are single planes of RSTU-VWXYZ Stars (top) and RSTUVWXYZ Rectangles (bottom).

For these two models
$\theta = 360/(9 - 1) = 360/8 = 45°$.

Since this angle can be obtained by the simple origami method of folding a right angle into half, no proof will be shown.

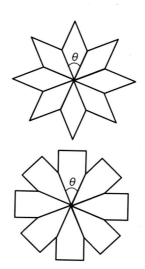

QRSTUVWXYZ Stars

Required angle of a unit at the center of each plane $\theta = 360/(10 - 1) = 360/9 = 40°$. So, required φ as marked in diagram $= \theta/2 = 20°$. Let the side of the starting square be of length a.

Fold a unit as shown on page 130 up to Step 4 and extend the new crease all the way to the edge at C. Before unfolding, i.e., when B lies on Q, trace the lines OQ and QC along the edges of the folded triangle. Hence, $\triangle OCQ \cong \triangle OCB$ as one is traced from the other.

In $\triangle OPQ$, $OQ = a$ and half diagonal $OP = a/\sqrt{2}$, and
$LOPQ = 180 - LAPQ$
$= 180 - 22.5 = 157.5°$.
Applying simple principles of trigonometry, $OP/\sin(LOQP) = OQ/\sin(LOPQ)$,
or $(a/\sqrt{2})/\sin(LOQP) = a/\sin 157.5$,
or $LOQP = \sin^{-1}((\sin 157.5)/\sqrt{2}) = 15.69°$.
Hence, $\beta = 180 - (157.5 + 15.69) = 6.81°$.

$LQOB = 45 + \beta = 45 + 6.81 = 51.81°$.
Since $\triangle OCQ \cong \triangle OCB$, $LQOC = LBOC$
$= LQOB/2 = 51.81/2 = 25.91°$.
Therefore, $\varphi = LQOC - \beta = 25.91 - 6.81$
$= 19.1° \approx 20°$, a close enough approximation for this model.

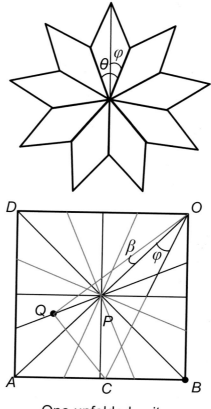

One of the ten planes.

One unfolded unit.

Afterword

Exercises

1. The figure on the right shows the creases for the traditional method of folding a square sheet of paper into thirds as presented in Chapter 1. *ABCD* is the starting square—*E* and *F* are midpoints of *AB* and *CD*, respectively. The line *GH* runs through the intersection point of *AC* and *DE* parallel to *AB*. Prove that *AG* is *AD*/3.

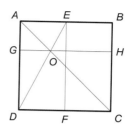

2. On page 24 there is a method showing how to obtain a pentagon from a square piece of paper. The figure on the right shows the original square and the resulting pentagon. Prove that the pentagon arrived at is close enough to a regular pentagon. If the starting square has a side length *a*, determine the side length of the resulting pentagon in terms of *a*. The shaded part shows the original square. (Answer: $a \sin 36°$)

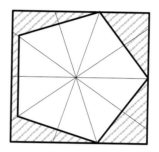

3. For Blintz Base Models without Inserts on page 43, try out the various colorings of the different flowers in a model as illustrated at the end but not diagrammed. Hint: Reverse some blintz folds from mountains to valleys.

4. The figure on the right shows the repeated blintz folding used in the Blintz Base Bouquets chapter. The starting square is S_0. When blintz folded once, the resulting square is S_1, when blintz folded again the resulting square is S_2 and so on until blintz folded *n* times resulting in the square S_n. If the length of a side of the starting square is *a*, prove that the length of a side of the *n*th square is $a/(\sqrt{2})^n$.

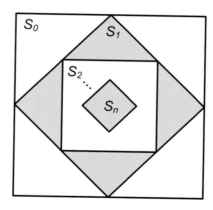

5. The Patterned Icosahedron model on page 58 is based on Lewis Simon and Bennett Arnstein's Triangle Edge Module [Gur95]. Pattern 1 on the right illustrates one finished face of the original Triangle Edge Module model. Pattern 2 illustrates one finished face of the Patterned Icosahedron in this book. Make minor variations to the Patterned Icosahedron unit to arrive at the finished faces shown in Patterns 3 and 4.

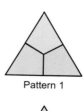
Pattern 1

Pattern 2

Pattern 3

Pattern 4

6. On page 74 there are diagrams showing how to obtain a rectangle of aspect ratio 1:(1+√3). Prove that the rectangle obtained by the folding method used is indeed of the said aspect ratio.

7. For the last five models in the Embellished Floral Balls chapter that starts on page 83 (Layered Poinsettia through Dianthus), find a way to employ inserts to simulate duo paper. Derive the insert size from the size of the rectangle used for the original unit. (Hint: For the Layered Poinsettia model on page 91, the insert may be introduced in Step 5.)

8. The figure below shows a flattened generic Embellished Floral Ball unit from Chapter 6. Points A and B of the unit end up at the center of the assembled flowers. Prove that in the folding method used, the angles marked θ are 60° angles.

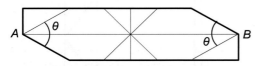

A generic Embellished Floral Ball unit.

9. Prove that for the planar model TUVWXYZ Hexagons on page 119, the folding method used to obtain the desired rectangle from a square is indeed of ratio 2:√3. Also prove that each assembled plane is a hexagon.

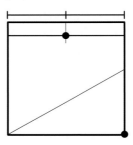

10. For the STUVWXYZ Heptagons model on page 125, prove that each assembled plane is approximately a heptagon.

Bibliography and Suggested Reading

◆ [Bee01] Rick Beech, *Origami: The Complete Practical Guide to the Ancient Art of Paperfolding*, Lorenz Books, 2001.

◆ [Bos01] *British Origami Society, BOS Magazine 208, June 2001.*

◆ [Bos03] *British Origami Society, BOS Convention 2003 Autumn Collection (+CD)*, 2003.

◆ [Cec89] Donatella Cecconi, *Origami Modulari 1* (in Italian), Il Castello, 1989.

◆ [Cecc89] Donatella Cecconi, *Origami Modulari 2* (in Italian), Il Castello, 1989.

◆ [Cox73] H. S. M. Coxeter, *Regular Polytopes*, Reprinted by Dover Publications, 1973.

◆ [Dir97] Alexandra Dirk, *Origami Boxes for Gifts, Treasures and Trifles*, Sterling, 1997.

◆ [Fer07] Bruno Ferraz, *Ultrapassando Fronteiras com o Origami (Exceeding Borders with Origami)* (in Portugese), Editora Ciência Moderna, 2007.

◆ [Fus89] Tomoko Fuse, *Origami Boxes*, Japan Publications Trading, 1989.

◆ [Fus90] Tomoko Fuse, *Unit Origami: Multidimensional Transformations*, Japan Publications, 1990.

◆ [Fus96] Tomoko Fuse, *Joyful Origami Boxes*, Japan Publications Trading, 1996.

◆ [Fus98] Tomoko Fuse, *Fabulous Origami Boxes*, Japan Publications Trading, 1998.

◆ [Fus00] Tomoko Fuse, *Quick and Easy Origami Boxes*, Japan Publications Trading, 2000.

◆ [Fus02] Tomoko Fuse, *Kusudama Origami*, Japan Publications Trading, 2002.

◆ [Fus06] Tomoko Fuse, *Unit Polyhedron Origami*, Japan Publications Trading, 2006.

◆ [Fus07] Tomoko Fuse, *Floral Origami Globes*, Japan Publications Trading, 2007.

◆ [Gil07] Eduardo Gil Moré, *Papiroflexia Y Geometría* (in Spanish), Miguel A Salvatella, 2007.

◆ [Gje08] Eric Gjerde, *Origami Tessellations: Awe-Inspiring Geometric Designs*, A K Peters, Ltd., to appear.

◆ [Gra76] Alice Gray, "On Modular Origami," *The Origamian vol. 13*, no. 3, page 4, June 1976.

◆ [Gur95] Rona Gurkewitz and Bennett Arnstein, *3-D Geometric Origami: Modular Polyhedra*, Dover Publications, 1995.

◆ [Gur99] Rona Gurkewitz, Bennett Arnstein, and Lewis Simon, *Modular Origami Polyhedra*, Dover Publications, 1999.

◆ [Gur03] Rona Gurkewitz and Bennett Arnstein, *Multimodular Origami Polyhedra*, Dover Publications, 2003.

◆ [Gur08] Rona Gurkewitz, *Beginner's Book of Multimodular Origami Polyhedra: The Platonic Solids,* Dover Publications, 2008.

◆ [Hul02] Thomas Hull, ed., *Origami 3: Third International Meeting of Origami Science, Mathematics, and Education*, A K Peters, Ltd., 2002.

◆ [Hul06] Thomas Hull, *Project Origami: Activities for Exploring Mathematics*, A K Peters, Ltd., 2006.

◆ [Jac87] Paul Jackson, *Encyclopedia of Origami/Papercraft Techniques*, Headline, 1987.

◆ [Joh66] Norman W. Johnson, "Convex Solids with Regular Faces", *Canadian Journal of Mathematics*, vol. 18, pages 169–200, 1966.

◆ [Kap01] Craig S. Kaplan and George W. Hart, "Symmetrohedra: Polyhedra from Symmetric Placement of Regular Polygons", in pages 21–28, Proceedings of Bridges 2001.

◆ [Kas98] Kunihiko Kasahara, *Origami for the Connoisseur*, Japan Publications, 1998.

◆ [Kasa98] Kunihiko Kasahara, *Origami Omnibus: Paper Folding for Everybody*, Japan Publications, 1998.

◆ [Kas03] Kunihiko Kasahara, *Extreme Origami*, Sterling, 2003.

[Kawai70] Toyoaki Kawai, *Origami*, Nursing, Inc., New Edition, 1970.

[Kaw02] Miyuki Kawamura, *Polyhedron Origami for Beginners*, Japan Publications, 2002.

[Kawa01] Toshikazu Kawasaki, *Origami Dream World* (in Japanese), Asahipress, 2001.

[Kawa05] Toshikazu Kawasaki, *Roses, Origami & Math*, Japan Publications Trading, 2005.

[Lan03] Robert Lang, *Origami Design Secrets: Mathematical Methods for an Ancient Art*, A K Peters, Ltd., 2003.

[Lan08] Robert Lang, ed., *Origami 4: Fourth International Meeting of Origami Science, Mathematics, and Education*, A K Peters, Ltd., to appear.

[Mit97] David Mitchell, *Mathematical Origami: Geometrical Shapes by Paper Folding*, Tarquin, 1997.

[Mit00] David Mitchell, *Paper Crystals: How to Make Enchanting Ornaments*, Water Trade, 2000.

[Muk07] Meenakshi Mukerji, *Marvelous Modular Origami*, A K Peters, Ltd., 2007.

[Ow96] Francis Ow, *Origami Hearts*, Japan Publications, 1996.

[Pet98] David Petty, *Origami Wreaths and Rings*, Aitoh, 1998.

[Pet02] David Petty, *Origami 1-2-3*, Sterling, 2002.

[Pet06] David Petty, *Origami A-B-C*, Sterling, 2006.

[Ran69] Samuel Randlett, *The Flapping Bird*, Samuel Randlett, 1969.

[Reh08] Julie J. Rehmeyer, "Math on Display", *Science News Online*, vol. 173, no. 7, Feb 2008.

[Rob04] Nick Robinson, *The Encyclopedia Of Origami*, Running Press, 2004.

[Row66] Tandalam Sundara Row, *Geometric Exercises in Paper Folding*, Reprinted by Dover Publications, 1966.

[Tak74] Toshie Takahama, *Creative Life with Creative Origami*, Volume I, Macaw Co., 1974.

[Tem86] Florence Temko, *Paper Pandas and Jumping Frogs*, China Books & Periodicals, 1986.

[Tem04] Florence Temko, *Origami Boxes and More*, Tuttle Publishing, 2004.

[Tub06] Arnold Tubis and Crystal Mills, *Unfolding Mathematics with Origami Boxes*, Key Curriculum Press, 2006.

[Tub07] Arnold Tubis and Crystal Mills, *Fun with Folded Fabric Boxes*, C&T Publishing 2007.

[Yam90] Makoto Yamaguchi, *Kusudama Ball Origami*, Japan Publications, 1990.

Suggested Websites

- [Aha] Gilad Aharoni, *Gilad's Origami Page*, http://www.giladorigami.com/

- [And] Eric Andersen, *paperfolding.com*, http://www.paperfolding.com/

- [Bos] British Origami Society, *BOS Home Page*, http://britishorigami.info/

- [Bur] Krystyna Burczyk, *Krystyna Burczyk's Origami*, http://www1.zetosa.com.pl/~burczyk/origami/

- [Har] George Hart, *The Pavilion of Polyhedreality*, http://www.georgehart.com/pavilion.html

- [Hul] Tom Hull, *Tom Hull's Home Page*, http://www.merrimack.edu/~thull/

- [Kat] Rachel Katz, *Origami with Rachel Katz*, http://www.origamiwithrachelkatz.com

- [Lan] Robert Lang, *Robert J. Lang Origami*, http://www.langorigami.com/

- [Muk] Meenakshi Mukerji, *Origami—MM's Modular Mania*, http://www.origamee.net/

- [Ori] Origami Resource Center, *Origami: the Art of Paper Folding*, http://www.origami-resource-center.com/index.html

- [Ous] Origami-USA, *Welcome to OrigamiUSA!*, http://origami-usa.org/

- [Ow] Francis Ow, *Francis Ow's Origami Page*, http://web.singnet.com.sg/~owrigami

- [Ped] Mette Pederson, *Mette Units*, http://mette.pederson.com

- [Pet] David Petty, *Dave's Origami Emporium*, http://members.aol.com/ukpetd/

- [Pla] Jim Plank, *Jim Plank's Origami Page (Modular)*, http://www.cs.utk.edu/~plank/plank/origami

- [Rez] Jorge Rezende, *Jorge Rezende home page*, http://gfm.cii.fc.ul.pt/Members/JR.en.html

- [Ros] Halina Rosciszewska-Narloch, *Haligami World*, http://www.origami.friko.pl

- [Sha] Rosana Shapiro, *Modular Origami*, http://www.ulitka.net/origami/

- [Shu] Yuri & Katrin Shumakov, *Oriland*, http://www.oriland.com

- [Tem] Florence Temko, *Origami*, http://www.bloominxpressions.com/origami.htm

- [Ter] Nicolas Terry, *Passion Origami.com*, http://www.passionorigami.com/

- [Tsu] Mio Tsugawa, *The Kusudama*, http://puupuu.gozaru.jp/

- [Ver] Helena Verrill, *Origami*, http://www.math.lsu.edu/~verrill/origami/

- [Vers] Paula Versnick, *Orihouse*, http://www.orihouse.com/

- [Wal] Dennis Walker, *Origami Database*, http://db.origami.com/

- [Wol] Wolfram Research, *Origami— from Wolfram MathWorld*, http://mathworld.wolfram.com/Origami.html

- [Wu] Joseph Wu, *Joseph Wu Origami*, http://www.origami.vancouver.bc.ca/

About The Author

Meenakshi Mukerji (Adhikari) was introduced to origami in early childhood by her metallurgist maternal uncle Bireshwar Mukhopadhyay. She rediscovered origami in its modular form as an adult, quite by chance in 1995, when she was living in Pittsburgh, Pennsylvania. A friend, Shobha Prabakar, took her to a class taught by Doug Philips, and ever since she has been folding modular origami and displaying it on her very popular website, http://www.origamee.net. She has many designs to her credit. In 2005, Origami USA presented her with the Florence Temko award for generously sharing her modular origami work. In April 2007, A K Peters Ltd published her first book, *Marvelous Modular Origami,* which became an origami bestseller within the year.

Meenakshi was born and raised in Kolkata, India. She obtained her BS in electrical engineering at the prestigious Indian Institute of Technology, Kharagpur, and then came to the United States to pursue a master's degree in computer science at Portland State University, Oregon. After successful completion of her studies, she joined the software industry and worked for more than a decade. She is now at home in California with her husband and two sons to enrich their lives, to create her own origami designs, and to author origami books. People who have provided her with much origami inspiration and encouragement are Rosalinda Sanchez, David Petty, Francis Ow, Rona Gurkewitz, Robert Lang, Rachel Katz, Ravi Apte, and the numerous viewers of her website.

Other Sightings of Author's Works

◈ *The San Jose Mercury News* and the community newspapers *The Cupertino Courier* and *The Sunnyvale Sun* featured an article about the author's work titled "Ancient Japanese Art of Origami is Growing in Popularity," in their May 21, 2008 issues.

◈ *Exhibition of Mathematical Art Catalog,* American Mathematical Society and Mathematical Association of America Joint Mathematics Meetings, San Diego, CA, 2008: author's Poinsettia Floral Ball photo was published and was also featured on the catalog cover.

◈ *Marvelous Modular Origami*: the author's first book was published by A K Peters, Ltd., in 2007.

◈ *Meenakshi's Modular Mania,* (http://www.origamee. net): Maintained by the author for the past decade or so, the website features photo galleries and diagrams of her own works as well as others' works.

◈ *Papiroflexia Y Geometría* by Eduardo Gil Moré, published by Miguel A Salvatella, Spain, 2007: The author's Plain Cube unit and adaptations were published.

◈ *Ultrapassando Fronteiras com o Origami (Exceeding Borders with Origami)* by Bruno Ferraz, published by Editora Ciência Moderna, Brazil, 2007: author's Whirl Cube model was published.

◈ Model Collection, BOS 40th Anniversary Convention 2007, British Origami Society: Diagrams of the author's Flowered Sonobe were published on CD.

◈ *Model Collection, Bristol Convention 2006*, British Origami Society: Diagrams of the author's Star Windows were published on CD.

◈ *The Encyclopedia of Origami* by Nick Robinson, Running Press, 2004: A full page photo of the author's QRSTUVWXYZ Stars model appears on page 131.

◈ *Reader's Digest*, June 2004 issue, Australia Edition: A photo of the author's QRSTUVWXYZ Stars model appears on page 17.

◈ *Reader's Digest*, June 2004 issue, New Zealand Edition: A photo of the author's QRSTUVWXYZ Stars model appears on page 15.

◈ *Quadrato Magico #71*, August 2003 issue (a publication of the Italian Origami Society, Centro Diffusione Origami): Diagrams of the author's Primrose Floral Ball appear on page 56.

◈ *Dave's Origami Emporium* (http://members.aol.com/ukpetd/): This website by David Petty features the author's Planar Series diagrams in the Special Guests section (May–Aug 2003).

◈ *Scaffold,* Vol. 1 Issue 3, April 2000: Diagrams of the author's Thatch Cube model appear on page 4.